SCIENTIFIC AND EDUCATIONAL JOURNAL OF THERAPEUTIC RIDING

2014

ISSN 1174-6238

ISBN 978-0-9845490-8-5

A PUBLICATION OF

THE FEDERATION OF HORSES IN EDUCATION AND THERAPY INTERNATIONAL A.I.S.B.L.

The Federation of Horses in Education and Therapy International A.I.S.B.L

SECRETARIAT
HETI
P.O.Box 134
Damariscotta, ME 04543
Tel: +44 207 1831 256
Email: office@HETIfederation.org
www.HETIfederation.org

ISSN 1174-6238
ISBN 978-0-9845490-8-5

Table of Contents

Dear HETI Members

It is a pleasure to greet the New Year with a new issue of the Scientific and Educational Journal of Therapeutic Riding (SEJThR). This issue of our Journal marks the completion of its transition from a simple compendium of articles to a true peer-reviewed scientific journal. This fulfills a goal set by our Executive Committee and we hope you, along with the members of the committee, will share our pride and excitement in achieving this milestone.

It has long been HETI's desire to support developments in equine therapeutic research and to give authors a forum that satisfies academic requirements while at the same time making them eligible to receive grants to continue and expand their research. An important prerequisite has been satisfied in attaining this benchmark.

Praise is in order for our two co-editors, Nina Ekholm Fry and Susan Cotton, who had to first establish submission protocols and manuscript standards before they could issue the call for papers, receive them and begin the editing process. It has been an arduous process but well done!

This wonderful news takes place against a backdrop of change for not-for-profit membership organizations which have seen a decline in their membership rolls due to the proliferation of online information. HETI has not been immune to this phenomenon. As members, you need to ask yourselves if having a peer-reviewed Journal, which serves as a voice in the future of therapeutic horsemanship, is important to you. Your membership in HETI, along with your contribution to this Journal, will enable research in equine therapy to continue and flourish and gain the recognition in the scientific community which it deserves. This will enable the next step in our goal for this Journal, its inclusion in searchable, scientific databases. It all begins here with you. Your contributions, and those of your professional colleagues are vital to our future.

We look forward to seeing you at our XV Congress in Taiwan in June where these issues and others are bound to be topics of discussion. This discourse is welcomed; it will be a healthy and enriching experience for all concerned. Please read this Journal, use it as a resource, share it with your colleagues, and reference it often. Our best wishes and appreciation go to each one of you for your continued support.

Ann O'Shallie
President, HETI (The Federation of Horses in Education and Therapy International A.I.S.B.L.)

Gisela Rhodes
Executive Director, HETI (The Federation of Horses in Education and Therapy International A.I.S.B.L.)

Editorial

As the Scientific and Educational Journal of Therapeutic Riding (SEJThR) continues to evolve, we thought it might be useful to begin this issue with a reminder of the nature of the journal, and the changes that have taken place this year.

> *The SEJThR is a journal that is scholarly, international, and specifically focused on horses and other equines in therapy, rehabilitation, and education/learning settings.*
>
> *Scholarly and international:*
>
> This year the SEJThR implemented a double-blind peer-review system, which supports the necessary rigor of a scholarly journal. We are excited about the international body of reviewers that have joined this year and welcome more! Please visit the journal website at www.hetifederation.org/journal for more information about serving the journal, and the field of equine-assisted activities and therapies, as a reviewer.
>
> *Specifically focused on horses and other equines:*
>
> The SEJThR is, to our knowledge, the only scholarly journal that focuses specifically on horses and other equines in therapeutic and learning settings.

The core of the journal is original research: studies and systematic literature reviews. That being said, the purpose of the journal is also educational, and we encourage submission of case studies, as these directly support both, practitioners and researchers, as well as articles in the "Theory and Practice" section, which range from explorations of relevant topics to proposals for research protocols.

In order to conduct studies with sufficient numbers of participants from homogenous populations, we need to collaborate. The Scientific and Educational Journal of Therapeutic

Riding creates an excellent opportunity for international information sharing, and it is our hope that it serves as a platform for future studies and connections. With the publication of the article Equine-Related Treatments for Mental Disorders Lack Empirical Support by Anestis and colleagues in the Journal of Clinical Psychology this year (see Ann Kern's commentary in the HETI December newsletter), it is especially important to keep moving forward with the design of research protocols and with procurement of research funding.

In this issue:

- **Naidoo et al.** conducted a study on the effects of standard therapeutic riding on heart rate variability in the autism spectrum population;
- **Bassette et al.** conducted a study on the promising effects on academic engagement using an equine-based math curriculum;
- **Page** conducted a study on the social and emotional effects of therapeutic riding on the autism spectrum population as related to social communication, behavior, and sensory profile; and
- **Hirsch** presents a model of breast cancer rehabilitation, reports on its application with clients, and proposes a research study related to range-of-motion and self-efficacy in breast cancer survivors.

As always, please direct your emails with comments and questions to editorial@hetifederation.org.

Sincerely,

Nina Ekholm Fry (U.S.A./Finland) and Sue Maree Cotton (Australia)

Call for Manuscripts

The Federation of Horses in Education and Therapy International (HETI) publishes the Scientific and Educational Journal of Therapeutic Riding (SEJThR) annually. We are currently seeking submissions with the purpose of advancing knowledge in the fields of therapeutic riding and equine-assisted activities and therapies.

The SEJThR has several submission categories:

Original Research
An original research study conducted by the author

Review
A comprehensive or systematic review of research on a specific topic

Case Study
A presentation of a single comprehensive case study or multiple case studies

Theory and Practice
A discussion of a specific topic in the field

Letter to the Editor
Comments on articles published in the SEJThR

Manuscripts are reviewed through a double-blind peer-review system and submissions are welcome throughout the year.

For more information about manuscript requirements and the submission process, please visit www.hetifederation.org/journal.html.

Please contact journal editors Nina Ekholm Fry and Sue Maree Cotton with questions.

Journal Editors:
Susan Maree Cotton • Nina Ekholm Fry
editorial@hetifederation.org
www.hetifederation.org

Call for Reviewers

The Federation of Horses in Education and Therapy International (HETI) publishes the Scientific and Educational Journal of Therapeutic Riding (SEJThR) annually. The focus of this journal is on advancing knowledge in the fields of therapeutic riding and equine-assisted activities and therapies.

Reviewers are needed to ensure the quality and status of the SEJThR as a peer-reviewed journal. Peer-reviewers conduct blind reviews of manuscript submissions within their area of knowledge as a step in the review process.

Expectations for reviewers are:

- Ability to provide constructive feedback for a scholarly manuscript in terms of its quality and contribution to the field
- Timely and respectful communication with journal editors

Requirements for reviewers are:

- Professional experience in an area of therapeutic riding/equine-assisted activities and therapies
- A master's or doctoral degree
- Experience in reviewing scholarly texts
- Ability to complete a review within four weeks of receiving a manuscript

If you are interested in contributing your time as a peer-reviewer for the SEJThR, please submit a letter of interest and your curriculum vitae (CV) per email to editorial@hetifederation.org. Please type "Reviewer Application" in the subject line.

ARTICLE ONE

Acute Heart Rate Variability Responses to a Therapeutic Horseback Riding Session in Children with Autism Spectrum Disorders: A Pilot Study

Rowena Naidoo (PhD Sp Sc), Zingisa Z Nqwena (B Sp Sc Hon), Lauren Reimers (B Sp Sc Hon), Kate Peters (B Sp Sc Hon), Takshita Sookan (M Sp Sc) and Andrew J McKune (D Tech) - University of KwaZulu-Natal, Durban, South Africa, Discipline of Biokinetics, Exercise and Leisure Sciences

Correspondence to:
Rowena Naidoo
University of KwaZulu-Natal, School of Health Science
Private Bag X54001
Durban, 4000 South Africa
Email: naidoor3@ukzn.ac.za

Support:
The Ridegtop Equestrian Center, Riding for the Disabled, Durban, KwaZulu-Natal, South Africa

Acknowledgements:
The researchers would like to thank the children and families for participating in the study. We would also like to thank Mrs Tracey Cumming (SARDA Senior Instructor; BSocSc UKZN; SANEF 1) for the use of the facility, assistance with the THR sessions and her expertise in the field.

ACUTE HEART RATE VARIABILITY RESPONSES TO A THERAPEUTIC HORSEBACK RIDING SESSION IN CHILDREN WITH AUTISM SPECTRUM DISORDERS: A PILOT STUDY

ABSTRACT

Aim: The aim of this study was to determine the acute effects of a therapeutic horseback riding session (THR) on heart rate variability in children with autism spectrum disorders (ASD).

Methods: Heart rate variability time and frequency domain parameters were measured in five participants (6-17 years) with ASD. The heart rate variability parameters calculated were time domain: mean heart rate, inter-beat intervals and root mean squared differences of the standard deviation (RMSSD) and frequency domain: high frequency, low frequency and low frequency to high frequency ratio. Heart rate variability was recorded pre, during and post a 15-25 min THR session. These time point measurements were then compared using one way ANOVA with Tukey post hoc testing. Significance was set at $p < 0.05$.

Results: Despite no significant changes observed in heart rate or inter-beat intervals, an increase in RMSSD was observed post-THR ($p = 0.02$). Low frequency ($p = 0.0006$) and high frequency ($p = 0.0001$) parameters were significantly increased post versus pre and during THR, which suggested an increased activation of the autonomic nervous system. The post-THR low frequency to high frequency ratio was significantly reduced ($p = 0.04$) indicating elevated parasympathetic activity after THR compared to pre-THR values.

Conclusion: There is no conclusive evidence showing that participants with ASD demonstrate increased parasympathetic activity acutely after a session of THR. However, positively, the study suggests that it is possible to measure HRV in children with ASD in relation to THR.

KEYWORDS

Heart rate variability; Therapeutic Horseback Riding; Autism.

INTRODUCTION

Autism spectrum disorders (ASD) are part of a broad spectrum of neurodevelopmental disorders that occur in childhood. Children with ASD demonstrate impairments in social interaction, communication (verbal and nonverbal) and the presence of restricted and repetitive stereotyped behaviours (Ashwood, Wills, & Van de Water, 2006). Deficits in cognition and attention in children with autism have been linked to problems in the early stages of stimulus processing and to autonomic dysregulation, but the evidence is conflicting (Zahn, Rumsey, & Van Kammen, 1987).

Animal assisted therapy is defined as using an animal as a fundamental part of an individual's treatment process. Selected animal assisted therapy interventions involve the treatment of cognitive functioning and social interaction problems (Fine, 2010). Equine assisted activities include therapeutic horseback riding (THR), interactive vaulting and other horse-related activities. In contrast to hippotherapy which is performed by a professional therapist using equine movements for improvements in posture, balance and fine motor skills, THR is performed by a trained riding instructor teaching the disabled rider to control the horse using basic riding skills and incorporating activities to support the goals of the rider (Snider, Korner-Bitensky, Kammann, Warner, & Saleh, 2007). THR is defined as the use of the horse and equine oriented activities to achieve a variety of goals such as physical, emotion, social, cognitive, behavioural and educational (Lessick, Shinaver, Post, Rivera, & Lemon, 2004). Studies evaluating THR on individuals with physical disabilities, report statistically significant improvements in physical strength, tone and endurance; memory; behavioural control; and response to physical contact (King, 2007; MacKinnon, 2007; Uchiyama, Ohtani, & Ohta, 2011). Furthermore, THR is used to improve the riders posture, balance and coordination (Lessick, et al., 2004), while the riders vestibular system is also stimulated by the horse's change of direction and speed (Granados & Agís, 2011).

The mechanisms underlying the positive effects of THR are unclear. It has been suggested that the movement of the pelvis of a horse resembles human ambulation and this provides motor and sensory inputs similar to those received during human walking (Bowes and Cook, 2007). The region of contact between the rider's ischial bones and the horse's back is the centre of gravity. Motor pulses (beats) of the horse (three-dimensional sway) are transmitted to the rider through this region (the support). In turn, the rider perceives the pulses and responds with his or her own movements, which are transmitted back to the horse which results in a biofeedback mechanism (Uchiyama et al., 2011). Recent studies have shown that THR is beneficial for children with ASD (Gabriels et al., 2012; Uchiyama et al., 2011). Wuang et al. (2010) concluded that occupational therapy followed by a 20-week simulated developmental horse riding programme enhanced motor and sensory efficiency in children with ASD.

The physical benefits of THR have been documented, however, the psychophysiological benefits of THR is limited. Interacting with animals influences social interaction between humans and related factors important in this respect, such as trust, aggression and a positive mood (Beetz et al., 2012). A study by Bass et al. (2009) determined that THR resulted in less inattention and improved social motivation in children with ASD compared to their counterparts not participating in THR. Furthermore, THR stimulates multiple domains of functioning and therefore it may be applicable to assist with the treatment of neurological disorders.

Heart rate variability (HRV) is a reliable and non-invasive method for assessing the functioning of the autonomic nervous system (ANS), in particular vagal regulation (Acharya, Joseph, Kannathal, Lim, & Suri, 2006; Kleiger, Stein, & Bigger, 2005; Sookan & McKune, 2011; Taskforce, 1996). HRV reflects the changes in the interval between heart beats (R-waves) over time and it is largely dependent on the extrinsic regulation of the heart rate by the ANS. The time between one R-wave and the next, in milliseconds, is termed the RR interval or the inter-beat interval (IBI) (Algra, Tijssen, Pool, & Lubsen, 1993). The ANS controls the IBIs through the parasympathetic and sympathetic pathways (Sookan & McKune 2011). Parasympathetic system influences on the heart are often studied through measuring respiratory sinus arrhythmia (RSA) as an index of vagal activity. A larger RSA reflects greater vagal activity and a smaller RSA reflects lower vagal activity (Watson, Roberts, Baranek, Mandulak, & Dalton, 2012). High HRV is an adaptive quality of a healthy body, (Algra et al., 1993) whilst decreased HRV, which is an indicator of poor health, has been linked to a rise in sympathetic activity at rest (Lopes & White, 2006; Umetani, Singer, McCraty, & Atkinson, 1998).

Children with ASD have been shown to exhibit reduced levels of baseline RSA compared to controls (Bal et al., 2010; Ming, Julu, Brimacombe, Connor, & Daniels, 2005; Ming, Stein, et al., 2005; Van Hecke et al., 2009), as well as diminished adaptation in RSA during tasks with cognitive demands (Althaus et al., 2004; Toichi & Kamio, 2003). Ming et al. (2005) suggested that deficits in parasympathetic function among children with ASD may result in relatively unrestrained sympathetic activity. However, research examining HRV in children with autism is contradictory. Autistic children have been shown to have reduced parasympathetic activation (Ming et al., 2005a), as well as dysregulation of the sympathetic nervous system (Anderson, Colombo, and Unruh, 2013). However, Graveling and Brooke (1978) found that children with autism have an over active parasympathetic nervous system and a disproportionate sympathetic reaction to stimulation.

Therefore, based on the literature suggesting the benefits of THR on individuals with ASD and the conflicting research relating to ANS activation in individuals with ASD, the primary purpose of this study was to determine the acute effects of a THR session on HRV in children with ASD.

METHOD

PARTICIPANTS

A convenience sample of five (3 males and 2 females, 10.6 ± 4.2 years) children diagnosed with ASD participated in this study. All participants were diagnosed by physicians, however, severity was not specified for this study. All participants were capable of some, but very limited verbal communication. Participants were recruited from the Ridegtop Equestrian Centre, Riding for the Disabled, Durban in KwaZulu-Natal, South Africa where each had been participating in THR lessons for at least eight months.

TESTING PROCEDURES AND PROTOCOL

The Equestrian Centre provided the equipment, supplies, personnel and expertise at no cost to the researchers, as no funding was provided for this study. Additionally, the lessons for participants were free-of-charge. Ethical clearance (SHSEC 007/12) to conduct this study was granted by the University of KwaZulu-Natal research ethics committee.

Consent forms were signed by parents or guardians of each participant and procedures were explained, prior to testing, and on the day of testing. The protocol was performed over two days, one week apart. The first day was used as a familiarisation session. Measurements recorded from the second testing session were analysed. Additionally, two ECG electrodes (Unilect 4040M), with instructions on their placement, were provided to parents or guardians one week prior to the familiarisation day. The purpose of this was to ensure that participants were comfortable with electrodes placed on their chest for the two HRV measurement days. However, there was no report back from parents on the use of electrodes.

HEART RATE VARIABILITY MEASUREMENT

All THR sessions were performed individually and took place between 8-11 am on both testing days. Actiheart (Cambridge Neurotechnology, Cambridge, UK) monitors were used to record the inter-beat intervals (IBIs). The monitors are flat and lightweight (<10 g) devices which have high levels of intra- and inter-instrument reliability as well as good validity of measures (Brage, Brage, Franks, Ekelund, & Wareham, 2005; Kristiansen et al., 2011; Stalder, Evans, Hucklebridge, & Clow, 2011). This system is a suitable stand-alone ambulatory method and is the least invasive method of measuring HRV (Kristiansen et al., 2011).

Upon arrival at the Equestrian Centre two ECG electrodes (Unilect 4040M) were placed on the children below the sternum (V4 and V6). Subsequently, the Actiheart monitors were attached to the upper chest area via clips that fit onto the electrodes (Unomedical Ltd., Redditch, UK). The same researcher attached the electrodes together with the Actiheart monitor on all participants. A signal test was then performed using Actiheart 2.132 software (Cambridge Neurotechnology, Cambridge, UK).

For pre-THR and post-THR HRV measurement the children were requested to remain seated on a chair and be as still as possible for 3-5 minutes. Generally, children with ASD would find it difficult and stressful to remain in a single position for a prolonged period of time. Thus, to achieve this, the THR instructor tried as much as possible to minimise movement by interacting with the participant. Interactions included a question and answer session. For example, questions like "What is the colour of your clothing?" and "What is the name of the horse that you are riding?" Dialogue that participants found interesting was encouraged. After the pre-THR HRV measurement the children were placed on their horses and HRV was recorded for the duration of the THR session that varied from 15 to 25 minutes depending on each child.

Therapeutic Horseback Riding Lesson

The individual THR lesson was taught and led by a qualified THR instructor, together with two trained volunteers. The same two volunteers assisted with all sessions. One volunteer lead the horse, while the other was side walking, ensuring proper posture during riding by providing verbal instruction and/or demonstration if necessary. Participants did not require assistance/touch by the side walker during most activities with the exception when trotting. The participants were familiar with the instructor and the volunteers as they had been riding for at least eight months. Horses were individually selected per participant, depending on the participant's weight and height. Each participant was familiar with their horse, thus the same horse was ridden for each participant on both testing days. The THR sessions were conducted in a covered arena that is open on all sides.

Mounting/dismounting. Participants were assisted by the THR instructor to mount/dismount the horse. The mounting/dismounting activity lasted at least 3 minutes and was aimed at stimulating vestibular and sensory processing (Bass et al., 2009).

Riding. The riding session was individually designed to stimulate sensory processing, as well as gross and fine motor movements. Gross movements included catching and throwing a ball to the THR instructor, placing a hoop onto a specified point and turning around to touch the horse's back. Activities like selecting reins, adjusting how the reins are held and placing stickers onto the selected areas of the horse can increase fine motor skills. Participants were also instructed to perform the following skills: direct left and right rein changes of direction, upward and downward transitions (halt/walk/trot, trot/walk/halt) while performing activities specific to the participant's individual programme goal. Interaction with the side-walker during each participant's session was designed with the social interaction/communication goals specific to the participant. Programme goals for most participants were to improve their gait and posture, increase muscle tone as well as promote social interaction. Participants did not canter in the session.

Data Analysis

For HRV analysis, IBIs were recorded using the Actiheart software 2.132 and edited in a two-step procedure. Firstly data was edited using the built in automated editing programme, followed by manual inspection of the complete IBI time series and the removal of any remaining artifacts. The IBI's were then transferred to HRV Analysis 1.1 SPI software (Biomedical Signal Analysis Group, University of Kuopio, Finland) for the calculation of HRV time and frequency domains. For analysis of these HRV domains the software complies with the guidelines recommended by the Taskforce of the European Society of Cardiology and the North American Society of Pacing and Electrophysiology standards for measurement of HRV (Taskforce, 1996). The software analysed the raw IBI data and performed power spectral analysis using the autoregressive algorithm in accordance with the recommendations (Taskforce, 1996). The autoregressive algorithm was used as it yields improved resolution especially for short term HRV measurements (Roy, Boucher, & Comtois, 2009; Tarvainen & Niskanen, 2008). This algorithm creates a power spectral analysis with distinct frequency bands; namely high-frequency (HF), low-frequency (LF) and very low frequency (VLF). Sinnreich et al. (1998) suggested that the LF/HF ratio reflects sympathetic/parasympathetic balance. Chemoreceptor processes, thermoregulation, and the rennin-angiotensin system have been linked with the VLF band (Taskforce, 1996). The VLF range was not assessed in the present study. In addition, the time domains calculated by the software were mean heart rate, IBIs and root mean squared differences of the standard deviation (RMSSD) that has been shown to indicate vagal activation (Taskforce, 1996).

The time and frequency HRV domains were compared pre, during and post the horse riding session using a one-way analysis of variance and Tukey post-hoc testing was performed. Statistical analysis was performed using SPSS 19.0 software. Alpha level was set at $p < 0.05$. In order to avoid experimenter bias, the data was analysed by researchers that were not present during the testing days.

RESULTS

Frequency and time HRV domains pre, during and post THR are shown in Table 1. There were no significant changes in heart rate ($p = 0.24$) or IBI's ($p = 0.76$). When calculated as a percentage of their age-predicted heart rate maximum, the pre, during and post heart rates for the group ranged from 51.7 to 54.75% of heart rate maximum. These heart rates fell within the target heart rate zone (50-80% of maximal heart rate) recommended (American College of Sports Medicine, 2010) for developing cardiovascular fitness in apparently healthy individuals. However, taking into consideration that participants were part of a special population and that there were no differences between pre, during and post measurements, THR in this instance did not seem to be metabolically taxing for the children.

An increase in RMSSD was observed post-THR ($p = 0.02$), which suggested increased vagal activation, thus parasympathetic activity. Significant increases in LF ($p = 0.0006$) and HF ($p = 0.0001$) values were observed post-THR compared to during THR and pre-THR, suggesting an increased activation of the ANS. However, the post LF/HF ratio was significantly reduced ($p = 0.04$), suggesting elevated parasympathetic dominance (Table 1).

TABLE 1.

ANOVA RESULTS FOR FREQUENCY AND TIME HRV PARAMETERS PRE-, DURING AND POST- THR (MEAN (STANDARD DEVIATION))

HRV Parameter	Pre-THR	THR	Post-THR	p value
Time Domain				
Heart Rate (b/min)	112.9 (16.46)	107.3 (11.96)	114.2 (11.55)	0.24
Inter-beat Intervals (ms)	552.7 (69.14)	564.3 (62.45)	538.0 (58.77)	0.76
RMSSD	34.13 (26.68)	36.93 (26.03)	59.16 (34.72)a	0.023
Frequency Domain				
LF	1078 (825.5)	951.6 (546.3)	2489 (2201)b	0.0006
HF	445.3 (560.6)	455.8 (506.4)	1470 (1115)b	0.0001
LF/HF	3.67 (2.59)	3.08 (1.67)	1.94 (1.18)a	0.04

a= significant difference between pre and post-riding p< 0.05.

b= significant difference between pre and post-THR and THR and post-THR p<0 .05.

Discussion

The aim of this pilot study was to observe the acute effects of a THR session on HRV in children with ASD. This study provides preliminary evidence that THR activates the ANS in children with ASD with a greater parasympathetic activation relative to sympathetic activation post-riding compared to pre-riding.

Despite the use of THR for ASD, the underlying physiological mechanisms are poorly understood with some contradictory evidence. It was previously shown that elevated parasympathetic tone at rest is representative of homeostasis whereas a disruption of homeostasis is reflected by a depressed parasympathetic tone and is associated with stress and stress vulnerability. This disruption is a characteristic of ASD (Porges, 1995).

The LF HRV band is influenced by baroreceptor mediated regulation of blood pressure and reflects predominantly sympathetic activity (Moodithaya & Avadhany, 2009). The mean LF value pre-THR suggested increased sympathetic activity dominance at rest possibly indicative of autonomic dysregulation that is a characteristic of ASD (or present in children with autism). This problem with stimulus processing affects attention and cognition in autistic children (Zahn, Rumsey, & Van Kammen, 1987). The reduction in the LF/HF ratio post-riding in the present study indicated a higher respiratory sinus arrhythmia, which has previously been shown to reflect a greater myelinated vagal control of the heart, and in turn suggest a soothed autonomic state (Patriquin, Scarpa, Friedman, & Porges, 2013). In addition, respiratory sinus arrhythmia is also positively associated with emotion regulation and other self-regulatory attributes that are crucial to effective social engagement (Porges, 2011). Therefore, it is possible that THR stimulates the parasympathetic nervous system and in turn may potentiate homeostasis within the ANS in children with ASD.

In addition, the RMSSD estimates the short term component of HRV and is an index of parasympathetic activation (Sookan & McKune, 2011). Participants in this study demonstrated a significant increase in RMSSD post-THR compared to pre-THR. This elevation suggested that the participants experienced a reduced amount of stress or arousal post-THR.

There were several limitations to this study. Firstly, no information regarding medication regimens was taken into consideration. The pre, during and post HRV recording times varied from 3-5 minutes. This was because it was difficult to control the behaviour of the participants, without causing extra stress/anxiety (Graveling & Brooke, 1978). In addition, other variables such as the weather, a change in routine, as well the presence of researchers on the day of testing may have had an effect on the arousal and or stress levels of the participants. This study did not examine the possible mechanisms that could explain why there would be increased parasympathetic activation post-THR. Finally, because of the small sample size in this pilot study, generalisations cannot be made.

Therefore, due to the limitations of the study, specifically the small sample size, there is no conclusive evidence showing that participants with ASD demonstrate increased parasympathetic activity acutely after a session of THR. However, the study suggests that it is possible to measure HRV in children with ASD in relation to THR.

Future research should extend to a larger population of children with autism and the chronic effects of THR on HRV and its relationship with behavioural, cognitive and health parameters should be examined. Additionally, a sample of age and gender matched normal functioning participants should be included as a control group to provide comparative data.

Conflict of Interest Statement:

The authors declare that the research was conducted in the absence of any commercial or financial relationships that could be construed as a potential conflict of interest.

REFERENCES

Acharya, U., Joseph, K., Kannathal, N., Lim, C., & Suri, J. (2006). Heart rate variability: a review. Medical and Biological Engineering and Computing, 44(12), 1031-1051. doi: 10.1007/s11517-006-0119-0

Algra, A., Tijssen, J. R., J, Pool, J., & Lubsen, J. (1993). Heart rate variability from 24-hour electrocardiography and the 2-year risk for sudden death. Circulation, 88, 180-185.

Althaus, M., Van Roon, A., Muler, L., Mulder, G., Aarnoudse, C., & Minderaa, R. (2004). Autonomic response patterns observed during the performance of an attention-demanding task in two groups of children with autistic-type difficulties in social adjustment. Psychophysiology, 41(6), 893-904. doi: 10.1111/j.1469-8986.2004.00252.x

American College of Sports Medicine. (2010). ACSM's Guidelines for Exercise Testing and Prescription (8th ed.). Philadelphia (PA): Lippincott Williams & Wilkins.

Anderson, C. J., Colombo, J., & Unruh, K. E. (2013). Pupil and salivary indicators of autonomic dysfunction in autism spectrum disorder. Developmental Psychobiology, 55(5), 465-482. doi: 10.1002/dev.21051

Ashwood, P., Wills, S., & Van de Water, J. (2006). The immune response in autism: a new frontier for autism research. Journal of Leukocyte Biology, 80(1), 1-15. doi: 10.1189/jlb.1205707

Bal, E., Harden, E., Lamb, D., Van Hecke, A., Denver, J., & Porges, S. (2010). Emotion Recognition in Children with Autism Spectrum Disorders: Relations to Eye Gaze and Autonomic State. Journal of Autism and Developmental Disorders, 40(3), 358-370. doi: 10.1007/s10803-009-0884-3

Bass, M. M., Duchowny, C. A., & Llabre, M. M. (2009). The effect of therapeutic horseback riding on social functioning in children with autism. Journal of Autism & Developmental Disorders, 39(9), 1261-1267. doi: http://dx.doi.org/10.1007/s10803-009-0734-3

Beetz, A., Uvnäs-Moberg, K., Julius, H., & Kotrschal, K. (2012). Psychosocial and psychophysiological effects of human-animal interactions: the possible role of oxytocin. Frontiers in Psychology, 3, 1-15. doi: 10.3389/fpsyg.2012.00234

Bowes, C., & Cook, B. (2007). The value of horse riding and hydrotherapy in the management of severe and complex disability. USA: Butterworth Heinemann, an imprint of Elsevier.

Brage, S., Brage, N., Franks, P., Ekelund, U., & Wareham, N. (2005). Reliability and validity of the combined heart rate and movement sensor Actiheart. European Journal of Clinical Nutrition, 59(4), 561-570.

Fine, A. (2010). Handbook on Animal-Assisted Therapy: Theoretical Foundations and Guidelines (3rd ed.). London: Academic Press.

Gabriels, R. L., Agnew, J. A., Holt, K. D., Shoffner, A., Pan, Z., Ruzzano, S., . . . Mesibov, G. (2012). Pilot study measuring the effects of therapeutic horseback riding on school-age children and adolescents with autism spectrum disorders. Research in Autism Spectrum Disorders, 6(2), 578-588. doi: 10.1016/j.rasd.2011.09.007

Granados, A., & Agís, I. (2011). Why children with special needs feel better with hippotherapy sessions: A conceptual review. Journal of Alternative & Complementary Medicine, 17(3), 191-197. doi: 10.1089/acm.2009.0229

Graveling, R. A., & Brooke, J. (1978). Hormonal and cardiac response of autistic children to changes in environmental stimulation. Journal of Autism and Developmental Disorders, 8, 441-455.

King, N. (2007). Perceived efficacy of therapeutic riding for children with autism. In B. Engel & J. MacKinnon (Eds.), Enhancing Human Occupation Through Hippotherapy (pp. 119-126). Bethesda, MD: American Occupational Therapy Association.

Kleiger, R. E., Stein, P. K., & Bigger, J. T. (2005). Heart rate variability: Measurement and clinical utility. Annals of Noninvasive Electrocardiology, 10(1), 88-101. doi: 10.1111/j.1542-474X.2005.10101.x

Kristiansen, J., Korshøj, M., Skotte, J., Jespersen, T., Søgaard, K., Mortensen, O., & Holtermann, A. (2011). Comparison of two systems for long-term heart rate variability monitoring in free-living conditions - A pilot study. BioMedical Engineering OnLine, 10(1), 27-40. doi: 10.1186/1475-925x-10-27

Lessick, M., Shinaver, R., Post, K., Rivera, J., & Lemon, B. (2004). Therapeutic horse riding: Exploring this alternative therapy for women with disabilities. AWHONN Lifelines, 8(1), 48-53.

Lopes, P., & White, J. (2006). Heart rate variability: Measurement methods and practical Implications. In P. J. Maud & C. Forster (Eds.), Physiological Assessment of Human Fitness Second Edition (pp. 39-62). Champaign, IL: Human Kinetics.

MacKinnon, J. (2007). What is the state of hippotherapy research and where should we go from here? In B. Engel & J. MacKinnon (Eds.), Enhancing Human Occupation Through Hippotherapy (pp. 158-173). Bethesda, MD: American Occupational Therapy Association.

Ming, X., Julu, P. O. O., Brimacombe, M., Connor, S., & Daniels, M. L. (2005). Reduced cardiac parasympathetic activity in children with autism. Brain & Development, 27(7), 509-516. doi: 10.1016/j.braindev.2005.01.003

Ming, X., Stein, T. P., Brimacombe, M., Johnson, W. G., Lambert, G. H., & Wagner, G. C. (2005). Increased excretion of a lipid peroxidation biomarker in autism. Prostaglandins, Leukotrienes and Essential Fatty Acids, 73(5), 379-384. doi: http://dx.doi.org/10.1016/j.plefa.2005.06.002

Moodithaya, S. S., & Avadhany, S. T. (2009). Comparison of cardiac autonomic activity between pre and post menopausal women using heart rate variability. *Indian Journal* of *Physiology* and *Pharmacology*, 53(3), 227-234.

Patriquin, M. A., Scarpa, A., Friedman, B. H., & Porges, S. W. (2013). Respiratory sinus arrhythmia: A marker for positive social functioning and receptive language skills in children with autism spectrum disorders. Developmental Psychobiology, 55(2), 101-112. doi: 10.1002/dev.21002

Porges, S. (1995). Cardiac vagal tone: A physiological index of stress. Neuroscience and Behavioral Reviews, 19(2), 225-233.

Porges, S. (2011). The Polyvagal Theory: Neurophysiological Foundations of Emotions, Attachment, Communication and Self-Regulation. New York, NY: W. W. Norton & Company.

Roy, A. R., Boucher, J. P., & Comtois, A. S. (2009). Heart rate variability modulation after manipulation in pain-free patients vs patients in pain. Journal of Manipulative and Physiological Therapeutics, 32(4), 277-287.

Snider, L., Korner-Bitensky, N., Kammann, C., Warner, S., & Saleh, M. (2007). Horseback riding as a therapy for children with cerebral palsy: Is there evidence of its effectiveness?. Physical and Occupational Therapy in Pediatrics, 27(2). doi: 10.1300/5006v27n02_02

Sookan, T., & McKune, A. (2011). Heartrate variability in physically active individuals: reliability and gender characteristics. Cardiovascular Journal of Africa, 22(3), 1-7.

Stalder, T., Evans, P., Hucklebridge, F., & Clow, A. (2011). Associations between the cortisol awakening response and heart rate variability. Psychoneuroendocrinology, 36(4), 454-462. doi: http://dx.doi.org/10.1016/j.psyneuen.2010.07.020

Tarvainen, M., & Niskanen, J. (2008). Kubio User Guide. Biosignal Analysis and Medical Imaging Group, University of Eastern Finland.

Taskforce. (1996). Heart rate variability: standards of measurement, physiological interpretation and clinical use. Task Force of the European Society of Cardiology and the North American Society of Pacing and Electrophysiology. Circulation, 93(5), 1043-1065.

Toichi, M., & Kamio, Y. (2003). Paradoxical autonomic response to mental tasks in autism. Journal of Autism and Developmental Disorders, 33(4), 417-426. doi: 10.1023/a:1025062812374

Uchiyama, H., Ohtani, N., & Ohta, M. (2011). Three-dimensional analysis of horse and human gaits in therapeutic riding. Applied Animal Behaviour Science, 135(4), 271-276. doi: http://dx.doi.org/10.1016/j.applanim.2011.10.024

Umetani, K., Singer, D. H., McCraty, M., & Atkinson, M. (1998). Twenty-four hour time domain heart rate variability and heart rate: relations to age and gender over nine decades. Journal of the American College of Cardiology, 31, 593-601.

Van Hecke, A. V., Lebow, J., Bal, E., Lamb, D., Harden, E., Kramer, A., . . . Porges, S. W. (2009). Electroencephalogram and heart rate regulation to familiar and unfamiliar people in children with autism spectrum disorders. Child Development, 80(4), 1118-1133. doi: 10.1111/j.1467-8624.2009.01320.x

Watson, L., Roberts, J., Baranek, G., Mandulak, K., & Dalton, J. (2012). Behavioral and physiological responses to child-directed speech of children with autism spectrum disorders or typical development. Journal of Autism and Developmental Disorders, 42(8), 1616-1629. doi: 10.1007/s10803-011-1401-z

Wuang, Y., Wang, C., Huang, M., & Su, C. (2010). The effectiveness of simulated developmental horse-riding program in children with autism. Adapted Physical Activity Quarterly, 27, 113-126.

Zahn, T., Rumsey, J., & Van Kammen, D. (1987). Autonomic nervous system activity in autistic, schizophrenic, and normal men: Effects of stimulus significance. Journal of Abnormal Psychology, 96(2), 135-144.

ARTICLE TWO

The Effects of an Equine-Based Mathematics Curriculum on Academic Engagement Behavior

Laura Bassette, Ph.D., BCBA-D, University of Illinois, Urbana-Champaign

Sarah Templin, M.Ed., Alan Beck, Ph.D., Mark Russell, Ph.D., Colleen Brady, Ph.D., Purdue University

Correspondence to:
Dr. Laura Bassette,
University of Illinois, Urbana-Champaign,
Department of Special Education, College of Education,
1310 S. 6th St.,
Champaign, Illinois 61820
Email: *labasset@illinois.edu*

THE EFFECTS OF AN EQUINE-BASED MATHEMATICS CURRICULUM ON ACADEMIC ENGAGEMENT BEHAVIOR

ABSTRACT

The purpose of this study was to measure academic engagement behavior in students with disabilities during an equine-based math curriculum compared to the traditional special education math curriculum. A repeated measures within subjects design was used to examine level of academic engagement and off-task behavior in six students with disabilities during an equine based functional math curriculum at an equine assisted activity center (i.e., curriculum that includes live horses and horse materials) compared to traditional mathematics instruction at school (i.e., no live horses or horse materials). Student behavior was recorded using the Behavioral Observation of Students in Schools (BOSS) code (Shapiro, 1996). Results indicate that students displayed a higher frequency of engagement during the equine-based curriculum. Areas for future research including effectively utilizing animals in special education curricula and the role of animal assisted interventions in learning are discussed.

KEYWORDS

Animal Assisted Interventions; Equine-Based Curriculum; Therapeutic Riding; Equine Assisted.

INTRODUCTION

Students with intellectual/developmental disabilities often receive services in self-contained settings (i.e., all students in the classroom are diagnosed with a disability) due to academic and/or behavioral difficulties (Heward, 2005; Westling & Fox, 2009). As a result of challenging behaviors, these students may experience academic difficulties, which impact their learning (Finn, Pannozzo, and Voelkl, 1995). Inappropriate behaviors are more often observed in students with intellectual disabilities compared to students without disabilities and may be a contributing factor to a child's academic difficulties. Behavioral challenges observed in students with disabilities can include: apparent lack of interest in learning, difficulty staying attentive in class, limited self-control, and aggressive behavior (Greenwood, Horton, and Utley, 2002; Heward, 2005; Westling & Fox, 2009). Furthermore these behaviors tend to lead to lower academic performance therefore limiting opportunities for academic engagement (Finn, Pannozzo, and Voelkl, 1995).

Academic engagement can be defined as behaviors such as writing, asking and answering questions, silently reading, and reading aloud (Stanley and Greenwood, 1983). Greenwood, Delquadri, and Hall (1984) noted that students with disabilities who struggle academically experience fewer opportunities to respond to academic material presented during lessons and are therefore less academically engaged. Furthermore, research demonstrated that the type of instruction provided can either promote or hinder academic engagement in students with disabilities; thus, exploring strategies to alter instruction and improve academic engagement for students with disabilities is warranted (Greenwood, Horton, & Utley, 2002).

One avenue to address the lack of engagement is to explore the impact of animal based instructional curriculum. Previous research found that animal-based curricula improve engagement behavior in students without disabilities (Witt, Ulmer, Burris, Brashers, & Burley, 2014) and students with disabilities (Katcher & Wilkins, 2000). The connection between children and animals is frequently observed and benefits such as attentiveness, motivation, and increased interest are anecdotally reported yet, little research exists which examines the reasons for this connection (Frisen, 2010; Melson, 2001). The idea of using animals through animal assisted activities for improving the learning of individuals who experience disabilities is an important area of further study (Melson, 2001).

For students without disabilities, Witt et al. (2014) found that students were more engaged during an inquiry-based animal agriculture curriculum compared to the traditional curriculum. Additionally, previous research demonstrates that one way to improve behavioral challenges and decreased academic engagement for children with disabilities may be through the use of animals (Bassette & Taber-Doughty, 2013; Katcher & Wilkins, 2000; Rud & Beck, 2000). Rud and Beck

(2000) conducted a survey of elementary schools teachers, which provided information on the presence and roles of companion animals in the classroom. Of the teachers that responded, 26.6% had animals in their classrooms and the types of animals reported by teachers to be in classrooms included fish, mammals, and reptiles as well as insects and birds. The teachers indicated pets were used to help teach academic subjects including creative writing and drawing assignments. Additionally, findings indicated guinea pigs helped to calm down children with disabilities during silent reading activities (Rud & Beck, 2000).

Katcher and Wilkins (2000) further examined the relationship between animals and children with disabilities by examining the effectiveness of their Companionable Zoo program. This program provided students with severe behavioral challenges an opportunity to learn how to take care of animals. The program involved teaching students how to be gentle and respectful towards animals. Following six months of participation, students demonstrated significantly lower incidences of severe behaviors in their regular school setting when compared to students in a control group. Reductions in aggression, social disruption, bullying, destruction of property, and impulsivity were also observed. Other behavioral changes included greater cooperation with instructors, greater engagement in learning, and more appropriate behavioral control in school. Overall, the students in this study responded to learning and demonstrated increases in appropriate behaviors when participating in the care of animals (Katcher & Wilkins, 2000).

More recently, Bassette and Taber-Doughty (2013) found academic engagement behaviors increased in elementary students with behavioral disabilities during a dog reading visitation program. These students were included in the study due to a lack of appropriate behavior during independent reading time. During intervention, students were able to read to a certified therapy dog and facilitator. Results found that students' engagement behavior increased from baseline indicating the dog was a source of motivation for appropriate behavior (Bassette & Taber-Doughty, 2013).

When considering adapting curriculum to incorporate animals, it is important to note that students with disabilities have the right to an appropriate educational program based on the Individuals with Disabilities Education Act (IDEA, 2004). For many students with intellectual disabilities, this is interpreted as access to a functional academic curricula focused on teaching the students life skills that will promote independence both in school and after formal schooling is completed, as the students enter adulthood (Westling & Fox, 2009). Functional mathematics is one area of functional academics that is a targeted area of instruction for students with intellectual disabilities. Skills taught under a functional mathematics curriculum may include: time management, measuring weight and height, understanding fractions for cooking, and money management (Westling & Fox, 2009). Typically, these skills are taught to students at school; however, embedding opportunities to learn these skills in other educational settings may promote generalization and maintenance, thus improving overall learning (Horn & Banerjee, 2009). Therapeutic horseback riding centers may serve as an ideal setting to address the individualized academic needs of students with disabilities through access to live animals.

While previous studies (Katcher & Wilkins 2000; Rud & Beck 2000) found that a variety of different animals may improve behavioral problems in the classroom, no study to date examined the impact of an equine-based functional mathematics curriculum taught at a therapeutic riding center on academic engagement behavior in children with disabilities. In the U.S., there are close to 850 facilities that are members of the Professional Association of Therapeutic Horsemanship International, one of the U.S. organizations that promote therapeutic horseback riding and interactions for people with a variety of disabilities (PATH Intl., 2014). Several recent reviews identified a variety of benefits of equine assisted activities for people with disabilities including physical benefits (Johnson, 2009); health benefits (Selby & Smith-Osbourne, 2013); and positive effects demonstrated in people with autism (O'Haire, 2013). These diverse areas demonstrate potential in the use of equine assisted activities; however, little empirical research is available that explores the effects of equine assisted activities on academic engagement behavior during academic tasks. Therefore, the purpose of the present study was to compare instruction that incorporated a horse-based functional mathematics curriculum vs. traditional mathematics curriculum (i.e., no live horses or materials), on academic engagement behavior in students with disabilities.

METHOD

PARTICIPANTS

Six students with disabilities were recruited for this study through a convenience sample from a special education class in the United States. After receiving Institutional Review Board (IRB) approval for the study, a purposive sample was used to recruit participants based on a recommendation by their special education teacher. Prior to the start of the study, parent/guardian(s) and students agreed to the terms of the study through written informed consent/assent. The class included eight students who met the study criteria of being able to verbalize and who were participating in an equine therapy program at a therapeutic horseback-riding center. The six students who were first to bring in their parental consent forms were selected as the targeted participants.

All six students were Caucasian, three male and three female. Two of the male participants were in 8th grade and one was in 6th grade. Two of the female participants were in 8th grade and one was in 7th grade. These students were taught all of their academic subjects in a self-contained classroom (all students in the classroom were diagnosed with a developmental disability) while other subjects (i.e., gym, art, and music) were taught in inclusive settings. Students qualified for special education serviced based on their IDEA diagnostic categories. Specifically, four students were identified as having a moderate intellectual disability (IQ<55), of which one also experienced an orthopedic impairment; the other two students were identified as having "other health impairments" and previous seizure disorders. Table 1 provides specific demographic information on each of the study participants.

TABLE 1: PARTICIPANTS' DEMOGRAPHIC INFORMATION

Student	Gender	Grade	Age	Disability	Prior participation in CHAPS program
1	Female	7th	13	Other Health Impairments, Previous seizure disorder	Yes
2	Male	8th	14	Moderate Mental Retardation	Yes
3	Female	8th	13	Moderate Mental Retardation	Yes
4	Female	8th	14	Moderate Mental Retardation and Orthopedic Impairment	Yes
5	Male	8th	14	Other Health Impairments, Previous seizure disorder	Yes
6	Male	6th	12	Moderate Mental Retardation	No

Setting

The special education classroom was a self-contained setting that served students with developmental disabilities (e.g., moderate intellectual disability). During the school observations, there were up to 16 students in the classroom along with the primary teacher and three paraprofessionals who were available to assist students with the mathematics lesson as needed. The student's primary teacher taught mathematics lessons at school during a thirty-minute class period.

The setting for the equine-based instruction was a therapeutic riding center called CHAPS. At the time of the study, CHAPS was a Professional Association for Therapeutic Horsemanship International premier accredited center where people with disabilities were given an opportunity to interact and learn through equine-related activities. Participants received both therapeutic riding lessons and classroom lessons at CHAPS once a week for the 8 weeks of the program. The students participated in the riding lesson in addition to the classroom component, however their academic engagement behaviors were only observed during the classroom lessons. A total of eight students participated in CHAPS' program. Six students served as participants in the study, and the other two students did not meet the inclusion criteria and were not included as participants. During each session, the class was divided in half, with one group riding then participating in the classroom, and the other group participating in the classroom, then riding. Therefore, there were four students in each of the classroom lessons at CHAPS. The classroom lessons at CHAPS were taught by a PATH International certified therapeutic riding instructor.

Experimental Design

A within subjects repeated measures design was used to determine if the equine mathematics curriculum would change the academic engagement behaviors of the students during the math classes. A within subjects design allows for each participant to serve as a control for themselves therefore each participant was observed at each level of the independent variables. This allowed for each of the conditions to be compared within all of the participants. A within subjects design is commonly used in research involving learning (Keren & Lewis, 1993). This design was chosen due to the associated advantages in obtaining a sufficient number of participants with a developmental disability (e.g., moderate intellectual disability) required for the experiment.

Dependent variable. The percentage of academic engagement behaviors as identified in the Behavioral Observation of Students in Schools (BOSS) code (Shapiro, 1996) was the dependent variable for the study. The BOSS code identified two categories of academic engagement including active engaged time (AET) and passive engaged time (PET). Active engaged time included behaviors such as writing, reading aloud, raising a hand, and talking to the teacher about the assigned material, while passive engaged time measured behaviors such as looking at the teacher, looking at an assigned worksheet, silently reading, looking at the blackboard, and listening to a peer talk about the assigned material. Additionally, total academic engagement was defined as the average of the active engaged time and the passive engaged time.

Independent variable. The horse curriculum was the independent variable for the study. The horse curriculum targeted four areas of the mathematics curriculum currently being taught in the self-contained classroom including fractions, measurement, money, and time/schedules. The objectives for each lesson were aligned with the alternative testing state standards for students who participated in an alternative educational curriculum. The fraction lessons consisted of writing whole and improper fractions through dividing horse related items such as carrots, apples, hay, water, etc. which were then fed to the horses afterwards. The measurement lessons consisted of measuring horses (height in hands and converting this to inches and feet) as well as horse related items such as saddles, girths, blankets, saddle pads, etc. The money lessons focused on calculating costs associated with caring for horses. Lastly, the time/schedule lessons consisted of reading a clock and using time to create a schedule for the day to take care of the horses. Lessons alternated between times with live horses and time with horse materials. During each classroom session at CHAPS, students also participated in a therapeutic horseback-riding lesson, however their behaviors were only observed during the classroom sessions, which did not include riding.

DATA COLLECTION

The research team consisted of three members, the author who served as the primary researcher and two additional data collectors. The data collectors assisted with video recording participants at the school and scoring reliability data. During observations, students were observed using a modified version of the Behavioral Observation of Students in Schools (BOSS) (Shapiro, 1996). The BOSS code was used to count the occurrences of academic engagement (active and passive time) during the students' math lessons. The two categories of academic engagement include either active engaged time (AET) or passive engaged time (PET).

The observation periods in the school classroom and at CHAPS were ten minutes and were divided into 40 intervals of 15 seconds in length. Every fifth interval, the teacher would be observed. During the intervals that the teachers were observed it was noted whether the teacher was directing the students, that is, presenting information to the students or working with a student one on one. If the teacher was talking with a teacher's aid, handing out papers, or doing any activity not directly involved with teaching the math lessons no note was taken.

For each ten-minute observation period, each student was observed for 32 intervals and the teacher was observed for 8 intervals. Students were observed and videotaped in both settings by the primary researcher at school and at CHAPS as well as by a second data collector in the school setting only. Tapes were reviewed at a later time and the level of academic engagement was recorded. A partial interval recording method was selected for the study. The partial interval recording method helped ensure any behaviors expressed within the entire interval were recorded whether they occurred at the beginning, middle, or end of the interval. Recordings were then reviewed and behaviors were recorded as either active or passive engaged behaviors. The total academic engagement time was then calculated by averaging the active and passive engaged behaviors and these calculations were used to analyze the effectiveness of the equine curriculum on academic engagement behaviors as compared to the school setting.

Procedure

Pre-training. An initial meeting was held with the students and the investigators prior to baseline data collection to allow students to become familiar with being observed and aimed to reduce any related stress. This phase also provided an opportunity for training the data collectors prior to baseline and intervention sessions.

Baseline. Baseline behavioral data were collected in the school classroom and at CHAPS for two weeks prior to intervention for a total of six baseline days at the school and three baseline days at CHAPS. The baseline period at CHAPS involved the students arriving at the equine facility and participating in math instruction led by a CHAPS instructor. During this phase, students were not exposed to any aspects of the horse curriculum (i.e. live horses or horse materials).

Intervention. The intervention period took place over an eight week time-frame during which students received the horse-related curriculum once per week at CHAPS from a PATH International certified therapeutic riding instructor. The classroom lessons for the eight-week program at CHAPS were divided into four weeks which used horse materials only (i.e., no live horses) and four weeks where live horses were included in the lessons. Concurrently, students were observed in their traditional special education classroom setting once per week.

Results

The percentage of time students were actively or passively engaged was calculated by dividing the total number of each category (active or passive) by the total number of intervals observed and multiplied by 100. Total time engaged was then calculated by averaging the active and passive engagement. Teacher Directed Instruction (TDI) time was also calculated by dividing the total number of occurrences by the total number of intervals the teacher was observed.

The totals for each category (active, passive, and total academic engagement) were analyzed. Analysis of variance (ANOVA) tests were conducted using Statistical Analysis Software (SAS) to determine statistical significance ($p<0.05$). Additionally, Tukey-Kramer adjustments were used for the pair-wise comparisons between each of the pairs of settings. The tests covered a comparison between the five time settings (six days of baseline at the school, three days of baseline at CHAPS, eight weeks of treatment at the school, four weeks of treatment with horse materials (but no live horses) at CHAPS, and four weeks of treatment with live horses at CHAPS to compare student behaviors. Other variables including student differences, order the students were observed during each observation, and time effects over week were also analyzed to determine any confounding variable effects.

The results indicate that the horse curriculum did increase academic engagement behaviors in participants. Specifically, there was an increase in the active academic engagement and total academic engagement during both the horse materials curriculum and live horse curriculum at CHAPS as compared to the school baseline and school treatment phases.

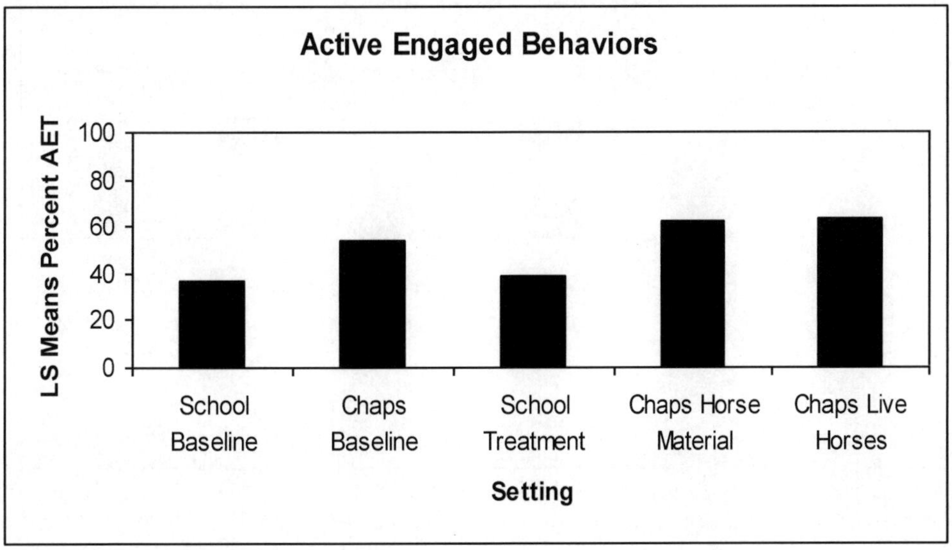

Figure 1: Mean percent of actively engaged academic behaviors in youth with disabilities using horse related materials at a therapeutic riding center vs. traditional school classroom.

Figure 1 illustrates the mean values for the active, engaged student behaviors during the baseline and intervention phases. Students demonstrated a higher rate of active engaged behaviors at CHAPS with the horse-related curriculum (p=0.0039) and when using live horses in the curriculum (p=0.0067) than during the school baseline period. Active engaged behaviors were also higher with the horse-related curriculum (p=0.0004), and when using live horses in the curriculum (p=0.0002) than during the school treatment period. There was no difference in active academically engaged behaviors between the horse-related curriculum, and the live horse curriculum.

No differences were found with passive engaged behaviors between the school classroom setting and the CHAPS with horse-related materials or with live horses. A higher frequency of passive engaged behaviors was observed during the CHAPS baseline as compared to the school baseline (p=0.0268). No difference in passive academically engaged behaviors were found between the horse related curriculum and the live horse curriculum.

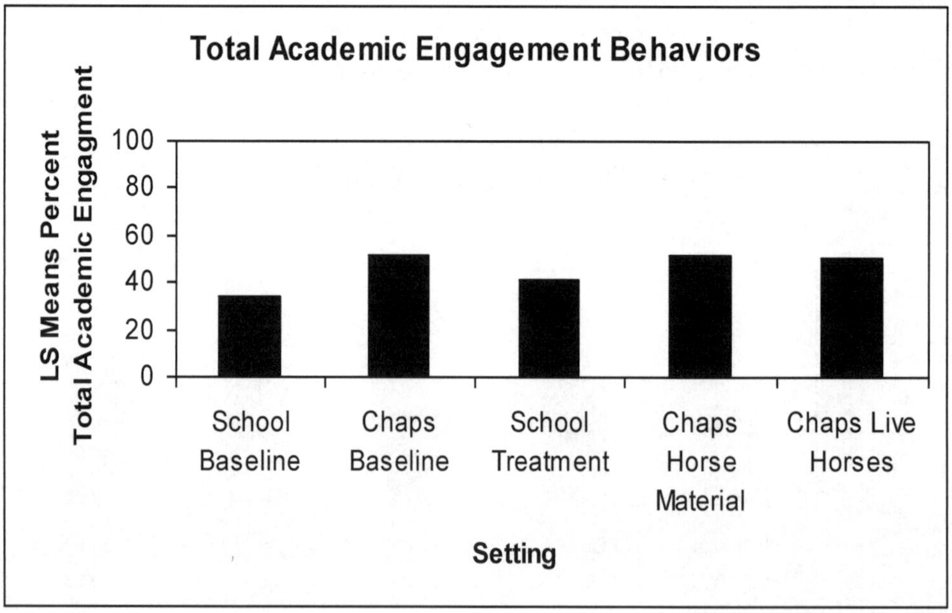

Figure 2: Mean percent of total engaged academic behaviors in youth with disabilities using horse related materials at a therapeutic riding center vs. traditional school classroom.

Figure 2 illustrates the mean values for the total academic engagement behaviors during the baseline and intervention phases. The total academic engagement behaviors were higher at

CHAPS with the horse related curriculum ($p<0.0001$), and when using live horses in the curriculum ($p=0.0007$) than during the school baseline period. Total academic engagement behaviors were also higher at CHAPS with the horse related curriculum ($p=0.0082$), and when using live horses in the curriculum ($p=0.0248$) than during the school treatment period. There was no difference in total academically engaged behaviors between the horse-related curriculum, and the live horse curriculum.

There was a significant effect on students in the passive behaviors ($p=0.0386$) and total engaged academic behaviors ($p<0.0001$), indicating individual student differences impacted their level of engagement. Specifically, for the passive engaged behaviors, students 3 and 4 had a significantly higher frequency of passive engaged behaviors than student 1 ($p=0.0333$ and $p=0.0104$) and student 4 also had a higher frequency of passive engaged behaviors than student 2 ($p=0.0398$). In the total academic engagement behaviors, student 3 had a higher frequency of engagement than student 1 ($p=0.0288$) and than student 2 ($p=0.0135$). Student 4 also had a higher rate of engagement than student 1 ($p=0.0176$), student 2 ($p=0.0079$), and student 6 ($p<0.0001$). While there were individual differences amongst the students in their engagement, there were no interactions between student differences and setting. This indicates that in both the school classroom and CHAPS' classroom the students were equivalently different in their levels of engagement, therefore the higher rates of academic engagement behaviors were caused by the horse curriculum not individual student differences.

INTER-OBSERVER AGREEMENT

Inter-observer agreement data were collected by a second member of the research team for 30% of the total number of observations. Percentage of occurrence and nonoccurrence agreement was calculated by dividing the number of intervals where there was agreement of the behavior either occurring or not occurring between both observers by the sum of the intervals where either one or both of the observers had scored the behavior as occurring or not occurring. This value was then multiplied by 100 and then reported as percentages. Mean percent of occurrence agreement for active engaged behaviors was 86.7% and for non-occurrence of active engaged behaviors the agreement was 89.7%. Mean percent of occurrence agreement for passive engaged behaviors was 85.0% and for non-occurrence of passive engaged behaviors the agreement was 82.6%.

DISCUSSION

The purpose of this study was to compare academic engagement behavior in students with disabilities in a horse-related functional math setting versus in a traditional functional mathematics setting. Data indicate that the equine-based mathematics curriculum taught at a therapeutic riding facility increased active academic engagement for middle school children with developmental disabilities. Effective methods of engaging students in the learning process is vital to academic success, and this form of differentiated instruction appears to have potential to promote behaviors consistent with increased learning. Since a lack of active engagement is related to students struggling in school even if students are on-task (Shapiro, 1996), the higher rates of active engagement during the horse related curriculum may lead to the students experiencing a higher level of academic progress and learning due to an increased number of opportunities to respond to the academic material and increased engagement in the learning process. The current study did not evaluate academic progress, but this area would be deserving of future assessment.

The only difference on the active engagement behaviors was based on the setting. There were no effects due to student differences, the order in which the students were observed, or in time. Additionally, the students' frequency of active engagement behaviors in this type of curriculum was the same whether they were observed at the beginning, middle, or end of the lesson.

LIMITATIONS AND RECOMMENDATIONS

Several limitations are apparent in this study. To begin, during intervention, there were four students in the classroom at CHAPS while at times there were up to 16 students in the school classroom; however, due to the paraprofessionals who assisted the teacher at the school, there were similar student/teacher ratios (e.g., a ratio of approximately 4-6 student to one teacher) across settings. Another limitation is a limited (3 day) baseline period at CHAPS all of which took place prior to the intervention period. There was an increase in the student's behaviors during the initial baseline period at CHAPS prior to the implementation of the equine-based curriculum indicating the change may have been a result of the change in setting, instructor, or class size and not the equine-based curriculum. Another limitation is that all of the students in this study participated in riding lessons while at CHAPS and this may have been a motivating factor for them during their classroom lessons. It would be of interest in future research to explore the role of motivation to interact with the horses during academic instruction.

The active engaged behaviors were higher in both the curriculum with horse-related materials and with the live horse portion than in the school classroom, indicating that either of these teaching methodologies was effective. It is interesting to note that there was not a difference in engagement between the use of live horses, and horse related materials. All but one student had previous exposure to live horses and this may have affected this student's behavior especially during the first couple of lessons. It would be interesting to explore if the horse related material effectively engaged students that had not had previous exposure to live horses. If that is true, the horse-related material could be used to engage students regardless of their ability to access a therapeutic riding center. Further studies with more students would strengthen the argument for the effectiveness of this instructional model. It would also be of interest to determine if the use of horse-related instructional materials in the traditional classroom setting increased active academic engagement.

An overall increase in academic engagement was observed during students' initial visit to CHAPS during the baseline period as compared to the school baseline period. This demonstrates that even though five of the students had previously participated in the program, there was still a potential novelty effect for the students and their behaviors were affected. The students may have been reacting to the change in setting and instructor and this impacted their behaviors. Additionally, there was no significant difference between the baseline period at CHAPS and either of the treatment periods. The baseline period for this study consisted of only three days during a two-week time frame prior to the implementation of the treatment period. Future research should seek to expand the amount of time the students participate in the equine curriculum to demonstrate if the increase in behaviors would be maintained once the novelty of the program has worn off.

Additionally, future studies should expand the baseline period or include a control to more specifically isolate the potential impact of the horse-related curriculum. Specifically, the effects of inter-mixing a control period, which uses lessons without horses within a time period with lessons that use the horse curriculum, should be examined. Doing this would provide more experimental control and more directly demonstrate the effect of the horse curriculum on the students' behaviors and help to eliminate any potential novelty effects.

By intermixing the interactions that the students have with the horse a more dramatic change in the behaviors may be observed. During the CHAPS baseline period the teacher counted down the days until the intervention phase when the students would be interacting with the horses (this also included when the riding lessons would start) and when the day came the students were very excited about being able to interact with the horses. Once the intervention phase started, the students understood that they would be interacting with the horses each time they came out and if this were not the case the students may have acted out or been less engaged. Future research that examines behavior and/or learning in students who are particularly attracted or indicate a preference towards horses prior to intervention would also be of interest (Bassette, 2014).

Educators who invest in developing an understanding of the effect of animals on students with disabilities may find a way to engage students in academic subjects. An understanding of the motivational factors and associated behavior that occur between humans and animals during animal assisted interventions is at its infancy. The impact of these interactions on behavior and learning is one that should be of interest particularly for students with disabilities who may not learn through traditional teaching strategies. This study provides a preliminary examination of the role of incorporating animals into academic curriculum for students with disabilities. Future research should seek to expand upon this study to help more thoroughly examine the best ways animal assisted interventions could be structured when addressing academic skills and/or behavior.

REFERENCES

Bassette, L. A. (May, 2014). Applying behavioral principles to human-animal interactions. Paper presentation at the meeting of the Association for Behavior Analysis International Annual Convention, Chicago, IL.

Bassette, L.A. & Taber-Doughty, T. (2013). The effects of a dog reading program on academic engagement behavior in elementary students with emotional and behavioral disabilities. Child and Youth Care Forum, 42, 239-246. DOI: 10.1007/s10566-013-9197-y

Finn, J. D., Pannozzo, G. M., & Voelkl, K. E. (1995). Disruptive and inattentive-withdrawn behavior and academic achievement among fourth graders. The Elementary School Journal, 95, 421-434.

Greenwood, C.R., Delquadri, J., & Hall, R. V. (1984). Opportunity to respond and student performance. In . Heward, T. Heron, D. Hill, & J. Trap-Porter (Eds.) Behavior in Education, (pp. 58-88). Columbus, OH: Merrill.

Greenwood, C.R., Horton, B.T., & Utley, C.A. (2002). Academic engagement: Current perspectives in research and practice. School Psychology Review, 31, 328-350..

Heward, William L. (2005). Exceptional Children: An Introduction to Special Education (8th ed.) Upper Saddle River, NJ: Pearson.

Horn, E., & Banerjee, R. (2009). Understanding curriculum modifications and embedded learning opportunities in the context of supporting all children's success. Language, Speech, and Hearing Services in Schools, 40, 406-415.

Individuals with Disabilities Education Improvement Act of 2004, 20 U.S.C. § 1400 et seq. (2004).

Johnson, C. C. (2009). The benefits of physical activity for youth with developmental disabilities: A systematic review. American Journal of Health Promotion, 23(3), 157-167.

Katcher, A. H. & Wilkins, G.G. (2000). The centaur's lessons: Therapeutic education through care of animals and nature study. In A. H. Fine (Ed) Handbook on Animal Assisted Therapy (pp. 153-177). New York, NY: Academic Press.

Keren, G. & Lewis, C. (1993). A Handbook for Data Analysis in Behavioral Sciences: Methodological Issues. Hillsdale, NJ: Lawrence Erlbaum Associates.

Melson, G.(2001). Why the Wild Things Are: Animals in the Lives of Children. Cambridge, MA: Harvard University Press.

O'Haire, M. E. (2013). Animal-assisted intervention for autism spectrum disorder: A systematic literature review. Journal of Autism and Developmental Disorders, 43, 1606-1622.

PATH International (2014). Retrieved from http://www.pathintl.org/

Rud, A. & Beck A.M. (2000). Kids and critters in class together. Phi Delta Kappan, 82, 313-315.

Scanlon, D., & Mellard, D. (2002). Academic and participation profiles of school age dropouts with and without disabilities. Exceptional Children, 68, 239-258.

Selby, A., & Smith-Osborne, A. (2013). A systematic review of effectiveness of complementary and adjunct therapies and interventions involving equines. Health Psychology, 32(4), 418-432.

Shapiro, E.S. (1996). Academic Skills Problems Workbook. New York: Guilford Press.

Stanley, S. & Greenwood C. (1983). How much opportunity to respond does the minority disadvantaged student receive in school? Exceptional Children, 49, 370-373.

Westling, D. & Fox, L. (2009). Teaching Students with Severe Disabilities. Upper Saddle River, NJ: Merrill.

Witt, P. A., Ulmer, J. D., Burris, S., Brashears, T., & Burley, H. A comparison of student engaged time in agriculture instruction. Journal of Agricultural Education, 55, 16-32.

ARTICLE THREE

The Social and Emotional Benefits of Therapeutic Riding for Children with Autism: An Exploratory Study

Cherié E. Page, Ph.D.

Correspondence to:
Dr. Cherié E. Page
Email: Cherie_e_page@yahoo.com

ABSTRACT

This quantitative exploratory study examined how therapeutic riding affects the emotional and social behaviors and symptoms of children aged 5 to 12, diagnosed with autism spectrum disorder (ASD). Parents of participants in the intervention group (n = 8) and comparison group (n = 5) completed pre- and post-intervention questionnaires. Participants in the intervention group received 15 weeks of continuous therapeutic riding, which consisted of one 30-minute session per week. Comparison group participants received a multidisciplinary educational approach. The guiding research question was: Does therapeutic riding have a beneficial effect on the social and emotional lives of children with a diagnosis of autism spectrum disorder, as assessed by the Social Communication Questionnaire (SCQ), the Child Behavior Checklist (CBCL), the Social Responsiveness Scale (SRS), and the Sensory Profile (SP)? Results revealed statistically significant positive changes pretest-posttest for the intervention group in social and emotional functioning in the forms of increased social communication, social interactions, reciprocal social skills, communication skills, and multisensory integration in social situations; and a decrease in autistic mannerisms, restricted and repetitive behaviors. Results for the comparison group also showed a significant increase in social and emotional coping strategies and a decrease in sensory-seeking behaviors to self-stimulate. The preliminary results from this study support evidence in the field that therapeutic riding can be an efficacious treatment for children with autism spectrum disorder. Due to the exploratory nature of this study, further studies are needed to evaluate the specific therapeutic effects of therapeutic riding has for children with autism.

KEYWORDS

Autism Spectrum Disorder; Communication Skills; Emotional Functioning; Multisensory Integration; Qualitative; Social Skills; Therapeutic Riding.

INTRODUCTION

Over the millennia, horses have arguably assisted humans in their evolution and currently horses are being worked with as a means for alternative therapies (All et al., 1999) and activities such as in therapeutic riding. Therapeutic riding is defined by the Horses and Humans Research Foundation (2010) as mounted activities including traditional riding disciplines or adaptive riding activities conducted by a certified instructor. Furthermore, the Professional Association of Therapeutic Horsemanship International (PATH, Intl.) states that therapeutic riding can positively contribute to the physical, social, emotional, and cognitive lives of individuals with specials needs. Therapeutic riding is commonly utilized to address the social and emotional impairments of children with autism spectrum disorder (Bass et al., 2009).

Autism spectrum disorder (ASD) can be defined as a disorder characterized by deficits in social communication and interaction, and patterns of behaviors (American Psychiatric Association [APA], 2013). Autism spectrum disorder is profound and comprises features that can include deficits in sensory integration, lack of social awareness and communication, inability to initiate directed attention, and restricted or stereotyped behaviors (Bass et al., 2009; Sams et al., 2006). It has been found to have an incidence rate of 1 in 110 live births and affects all populations and ages (Rosenzweig et al., 2002). While the cause of autism spectrum disorder is unknown, it is considered to be multifactorial (Bass et al., 2009), originating in a multifaceted combination of (a) genetics, (b) environmental factors, and (c) individual brain wiring (Gray, 2007).

The focus on possible neurobiological links (Keller & Persico, 2003; Roder, 2000) has been explored in past research. One possible hypothesis focuses on the possibility of an abnormality in genes that is central in the development of the brainstem. Studies have shown that the caudal part of the pons is smaller than normal in children with autism spectrum disorder (Rosenzweig et al., 2002). Additional research has shown there is a reduction in the corpus callosum as well as other cerebellar regions. Pierce and Courchesne (2001) reported that although a variety of neuroanatomical structures have been proposed, cerebellar abnormalities have been identified as a hallmark feature. This is supported by Bauman and Kemper (2005) who proposed that the cerebellum is involved in emotional and social functioning, which is consistent with some of the classic features of autism spectrum disorder. Given that therapeutic riding demands motor learning skills and social engagement, it can be a plausible means to affect change in the client who has been diagnosed with autism spectrum disorder.

A review of the extant literature revealed research that primarily consists of descriptive and observational reports (Bass et al., 2009; Biery, 1985; Bliss, 1997; Haehl, Giuliani, & Lewis, 1999;

Kaiser et al., 2004; Land et al., 2001; MacKinnon et al., 1995; Pauw, 2000). A study by Bass et al. (2009) was one of the first to explore the effects of therapeutic riding on the social functioning of children, age's five to 10, with autism spectrum disorder. It was hypothesized that 19 participants in the experimental group who received 12-weeks of therapeutic riding, compared to 15 participants in the waitlist control group, would experience improvements in social functioning. The social Responsiveness Scale and the Sensory Profile were utilized, to assess social functioning pre- and post-intervention. Results revealed children in the experimental group improved in critical areas, such as, directed attention, sensory integration, sensory sensitivity, decreased distractibility and inattention, and social motivation when compared to the control group. Bass et al. (2009) concluded that "the results of this study suggest that therapeutic horseback riding may be an efficacious therapeutic option for children with autism spectrum disorder" (p. 1266).

Research conducted since also supports the theory that therapeutic riding can improve the social lives of children with autism spectrum disorder. Ward et al. (2013) investigated the association between therapeutic riding and the social communication and sensory processing skills of children diagnosed with autism spectrum disorder. An interrupted treatment design was utilized to determine if 21 children who participated in a school group maintained the effects of therapeutic riding after the treatment was withdrawn as indicated by teachers' ratings of the children's behavior. The Gilliam Autism Rating Scale -2 (GARS-2) (Gilliam, 2006) and the Sensory Profile School Companion (SPSC) (Dunn, 2006) revealed that, following the intervention, social interactions and sensory processing significantly increased and the severity of symptoms associated with autism spectrum disorder decreased. Specifically, participants showed improvements in tolerance, attention, and responses to sensory input in the classroom. In addition, participants' capacity to relate in an appropriate manner to people, events, and objects significantly improved. While these benefits were not maintained when evaluation occurred two weeks later, they were regained during an additional 8-week therapeutic riding session; supporting evidence that consistency in therapeutic riding is necessary to maintain the benefits. Results of the study offer further evidence that therapeutic riding can be an effective intervention for children with autism spectrum disorder and that benefits witnessed can possibly transfer to classroom settings.

Kern et al. (2011) investigated the effects of equine-assisted activities on the severity of symptoms of 23 children with autism spectrum disorder. By utilizing the Childhood Autism Rating Scale (CARS) as well as the Timberlawn Parent-Child Interaction Scale on the quality of the parent-child interactions, children were evaluated at 3 to 6 months prior to the beginning of the waiting period, before starting therapeutic riding, and again after 3 and 6 months of riding. Results indicated that a reduction in the severity of symptoms occurred as a result of the therapeutic riding. The CARS pre-treatment baseline period showed no change; however, a significant decrease at the 3 and 6 month marker of riding was observed. The Timberlawn Parent-Child Interaction Scale revealed significant improvement in mood and tone at 3 and 6 months of riding. Additionally, marginal improvement in the reduction of negative regard at 6 months of riding was observed. Improvement was also seen within the parent-rated quality of life. These results suggest that children with autism spectrum disorder benefit from equine-assisted activities.

Lastly, Holm et al. (2014) examined how different doses of therapeutic riding impacted parent-nominated target behaviors for children with autism spectrum disorder. Three children, ages 6 – 8 years of age participated in a single subject multiple baseline, multiple case design, and received treatment doses of 1, 3, and 5 times per week. Target behaviors were collected and counted during the session, at home, and in the community. Compared to baseline, 70 % of the target behaviors improved during the intervention. Furthermore, these improvements were retained in 63 % of the behaviors during the withdrawal phase. It was discovered that increased doses of therapeutic riding were found to be significant for the degree of change to occur; and that these improvements were also able to be generalized to their home and community life.

While there is no known cure for autism spectrum disorder, or a consensus on the most appropriate treatment option, therapeutic riding may prove to be particularly effective. According to Bass et al. (2009), therapeutic riding provides a multisensory environment which "stimulates multiple domains and may be especially well-suited for children with neurological disorders who frequently present with a combination of motor, cognitive, and social disabilities" (p. 1261). Scarce empirically based research present within the field in combination with the poor rate for recovery and/or major improvement in children with autism spectrum disorder exposed to conventional psychotherapeutic and educational interventions (Freeman, Ritvo, Needleman, & Yokota, 1985; Meyer et al., 1987; Redfer & Goodman, 1989; Rutter, Bailey, & Lord, 2003) emphasizes the need to conduct research supporting the efficacy of this intervention.

The current study is a quantitative exploratory study examining how therapeutic riding affects the emotional and social behaviors and symptoms of 13 children with autism spectrum disorder. The hypothesis was that the children age's five to 12 with a diagnosis of mild to moderate autism spectrum disorder will benefit socially and/or emotionally from participating in a therapeutic riding intervention relative to base and relative to a non-treatment group, as measured by the Social Communication Questionnaire, the Child Behavior Checklist, the Social Responsiveness Scale, and the Sensory Profile.

METHOD

PARTICIPANTS

Eighteen participants were recruited, 2 dropped out before the study began, and 3 participants dropped out midway during the study. Thus, 13 participants completed the study, with 8 in the intervention group, and 5 in the comparison group (see Table 1). Inclusion criteria included chronological ages between 5 to 12 years, and a formal diagnosis of mild to moderate autism spectrum disorder by a physician in accordance with the DSM-IV-TR. Exclusion criteria included the inability of the child to meet the requirements needed to take part in therapeutic riding, including the ability to sit independently on the horse. Therapeutic riding was not the sole therapeutic intervention occurring, the following concurrently administered interventions were noted: occupational therapy (n = 12), speech therapy (n = 10), Applied Behavioral Analysis (n = 5), social skills groups (n = 5), adaptive physical education (n = 2), other behavioral therapies (n = 1), acupuncture (n = 1), Pivotal Response (n = 1), pressure therapy (n = 1), and gluten and/or casein free diet (n =1).

TABLE 1: PARTICIPANT CHARACTERISTICS

Characteristics	Intervention group (n = 8)	Comparison group (n = 5)
Sex		
Female	n = 2	n = 4
Male	n = 6	n = 1
Ethnicity		
Caucasian American	n = 6	n = 3
Mixed-race ethnicity	n = 1	n = 2
Asian Caucasian	n = 1	n = 0
Age	M = 7	M = 8.4
Classical Autism	n = 7	n = 3
Asperger's Syndrome	n = 1	n = 2
Parent sex	Female n = 7, Male n = 1	Female n = 5, Male n = 0

Participants for the intervention group were recruited from DreamPower Horsemanship, with one participant from One Step Closer, both centers providing therapeutic riding in Gilroy, California. Participants for the comparison group were recruited from El Torro Elementary in Morgan Hill; Autism Speaks Now, and Shelter Network. The comparison group received multidisciplinary education, including speech therapy, occupational therapy, social skills groups, and Applied Behavioral Analysis. Comparison group participants from El Torro Elementary took part in these therapies in the classroom weekly. For participants from Autism Speaks Now and Shelter Network, these therapies were occurring weekly within the typical lives of participants. Participant's involvement in therapeutic riding was also collected: new (n = 3), 1 year (n = 2), 1.5 years (n = 1), 2 years (n = 1), and 3 years (n = 1), as well as current and past medications and parent's own therapy/support groups. The goal was to recruit participants who had not taken part in therapeutic riding previously, however, due to time limitations of the recruitment phase this was not possible.

The equines that took part in the study were standard sized horses, breeds consisting of halflingers, quarter horses, Ponies of America, and paints. All equines were trained as therapeutic riding horses; had good temperament, tolerance, and a willingness to connect with humans. Equines ranged in ages from 12 to 20 years old and could accommodate riders up to 200 lbs.

Setting

The setting where the therapeutic riding lessons took place was inside a covered arena on the property where DreamPower Horsemanship was located. The covered arena had appropriate footing for the horses, a main entrance gate, as well as a side gate. The fence line was incorporated in some games and activities. The mounting block, along with poles, cones, cavalettes were kept on the outside of the arena in accordance with PATH Intl. standards, while the toys, rings, and beanie babies utilized were kept in a cabinet space outside the arena. Therapeutic riding instructors (TRI's) and/or the instructors-in-training were responsible for the set-up and tear-down of the arena. TRI's are certified by PATH Intl. TRI's in training were in the process of obtaining their certification, while under the supervision of a certified PATH Intl. TRI.

Measurements

The instruments were given to parents one week prior to the beginning of the intervention to establish a baseline, and again one week after the 15-week intervention ended.

Social Communication Questionnaire. The Social Communication Questionnaire (SCQ) was designed by Michael Rutter and Catherine Lord (Rutter et al., 2003). The questionnaire assesses social behavior, communication, and interaction within the last 3 months by utilizing yes-no parent responses to measure behaviors associated with autism spectrum disorder. The instrument has been validated by clinicians as containing both construct validity and reliability (Rutter et al., 2003). The primary standardization data for the SCQ came from a study containing 60 individuals with autism spectrum disorders and 40 individuals with non-autism diagnoses, but diagnoses of other language development disorders and mental retardation. Diagnostic validity was found by completing the following additional four steps:

a) item validity
b) receiver-operator characteristics,
c) factor analytic analysis, and
d) the correlation between the total SCQ and the autism diagnostic interview.

Child Behavior Checklist. The Child Behavior Checklist (CBC) was created by Thomas Achenbach (1991) and is designed to measure a child's behaviors and competencies. The questionnaire consists of 118, 3-point Likert-scale questions, as well as two open-ended questions that were added to describe additional issues. Likert scale ranges from 0 (*not true*) to 2 (*very true or often true*). The two open-ended questions were added by the researcher to assist in the qualitative data collection. The two open-ended questions were: *What concerns you most about your child?* and *Please describe the best thing about your child.* The measure was normed with 1,753 children between the ages of 6 to 18 years of ages in 48 states, and included various ethnic backgrounds. It was found that the measure contains both validity and reliability (Achenbach, 1991).

Social Responsiveness Scale The Social Responsiveness Scale (SRS), developed by Constantino (2002), is a 65-item questionnaire measuring the severity of autism spectrum disorder symptoms. It has a high internal consistency ($\alpha = .97$) and retest temporal stability in females and males ($r = .77$; $r = .85$, respectively). Each treatment scale was tested for internal consistency and it was found to have high Cronbach alpha scores, with social communication being the highest scored ($\alpha = .92$). The five treatment subscales measured by the SRS are social cognition, social awareness, social motivation, social communication, and autistic mannerisms. The SRS manual defines *social cognition* as the "ability to interpret social cues once they are picked up" (Constantino, 2002, p. 17), *social awareness* as the "ability to pick up on social cues" (p. 17), *social motivation* as "the extent to which a respondent is generally motivated to engage in social-interpersonal behavior" (p. 17), *social communication* as "expressive social communication" (p. 17), and *autistic mannerisms* as "stereotypical behaviors or highly restricted interests" (p. 17). The instrument was administered to parents who then rated their child on a 4-point Likert scale ranging from 0 (*not true*) to 4 (*almost always true*).

Sensory Profile The Sensory Profile (SP) was created by Dunn (1999), and is a 125-item questionnaire that is administered to parents. The instrument uses a 5-point Likert scale ranging from 1 *(always)* to 5 *(never)* and measures the overall social functioning as well as the degree to which children display problems along three categories:

a) sensory modulation,
b) sensory processing, and
c) behavioral and emotional responses.

The sensory modulation category includes subscales:

a) auditory processing,
b) visual processing,
c) vestibular processing,
d) touch processing,
e) multisensory processing, and
f) oral sensory.

The sensory-processing category includes subscales:

a) sensory related to tone,
b) modulation related to body position and movement,
c) modulation of movement affecting activity level,
d) modulation of sensory input affecting emotional regulation, and
e) modulation of visual input affecting emotional responses and activity level.

The behavioral and emotional category includes:

a) emotional and social response,
b) behavioral outcomes of sensory processing, and
c) items indicating thresholds for response.

PROCEDURES

The intervention occurred once a week, for 30 minutes, for 15 weeks at DreamPower Horsemanship, Gilroy. The child was assigned to the same horse, TRI, and side-walker each session. Issues such as size and weight of participant were taken into consideration when a horse was assigned. Uncontrollable events, such as illness of the participant lead to three sessions being rescheduled. One unforeseen and tragic event that occurred was the death of a therapeutic riding horse. This event was processed with the participant to the best of their developmental functioning. A new horse that was similar in size was chosen to replace the original horse. It was reported to researcher that several months after the loss of the participant's therapy horse, the participant would still indiscriminately inquire about where the horse was and request her to come to the session.

This study included children who were cognitively impaired for their age; several steps were taken to ensure consent. In all cases, parents signed the consent form for their child to participate. In some cases where the child was cognizant he or she was also allowed to sign a consent form agreeing to participate (see Appendices A and B).

Upon participant's arrival, they were fitted with a riding helmet and greeted by the TRI and/or TRI in training. Participants were assisted onto the mounting block inside the arena where they were assisted in mounting the horse. Participants were instructed to hold the reins, surcingle, or saddle pad while the side walkers on either side of the participant held onto the child in appropriate places, according to PATH Intl. standards. Participant was instructed to verbalize commands, such as "walk on," "please turn right/left," or "whoa," as a way to target social communication and awareness. During the 5 minute warm up, participants were asked to complete exercises such as reaching and pointing arms straight up, arm circles, trunk twists; and/or play "bunny and dragon". In this exercise, participants are asked to reach as far as they can to the horse's mane, and use their pointer and middle finger to make bunny ears, then as quickly as they can move the "bunny" to safety before the "dragon," played by the TRI, can capture the bunny. Participants then spent 15 minutes playing therapeutic games, such as tossing a ball to the TRI while the horse was in motion, tossing a ring onto a cone, playing red light/green light, letter and/or number games, and tossing beanie babies into specific buckets. These mounted games and exercises were designed to target and increase participants' social communication skills, social awareness, fine/gross motor skills, and perception coordination. Participants were then assisted with dismounting, in accordance with the PATH, Intl. standards.

DATA ANALYSIS

A paired *t*-test was utilized to examine the effect of the intervention of therapeutic riding for the quantitative data. A paired *t*-test can be defined as a way to compare means on the same participant over time. Due to the low *n* of the study, the paired *t*-test was conducted to explore inter-group comparisons. Results are indicative of the change over time that occurred for the intervention group and are not relative to the comparison group.

RESULTS

SOCIAL COMMUNICATION QUESTIONNAIRE

A paired *t*-test revealed significant pre and posttest differences within the intervention group, as shown in Table 2.

TABLE 2: SOCIAL COMMUNICATION QUESTIONNAIRE RESULTS

Total scale	Pre-test Mean	Post-test Mean	Δ Mean	*SD*	95% *CI*	Effect size	*p*
IG	17.00	13.63	3.38	3.62	[0.35, 6.40]	0.93	.034
CG	18.00	17.60	0.40	1.82	[-1.90, 2.66]	0.22	.648

Note. IG = intervention group, CG = comparison group. Effect Size is Cohen's *d*. 95%. Confidence Interval of the difference. *p* = Two-tailed.

Statistical significance was set at $p = <.05$ with $p = .034$. Significance was reached in regard to the intervention group on SCQ pretest and posttest scores. In addition, a large effect size was observed, $d = .93$.

Social Responsiveness Scale

A paired *t*-test revealed significant pre and posttest results for the intervention group, as well as with the subscales of social communication and autistic mannerisms. No significant differences within the intervention group for the following subscales: social awareness, social cognition, and social motivation, as evidenced by Table 3.

Table 3: Social Responsiveness Scale Results

Scales	Pretest Mean	Posttest Mean	Δ Mean	*SD*	95% *CI*	Effect size	*p*
Total scale:							
TG	84.88	78.88	6.00	5.07	[1.76, 10.24]	1.18	.012
CG	88.80	87.20	1.60	5.07	[-15.89, 19.08]	0.32	.812
Subscales:							
Social communication							
TG	81.88	74.63	7.25	4.62	[3.39, 11.11]	1.57	.003
CG	83.80	81.80	2.00	5.48	[-4.80, 8.80]	0.37	.460
Autistic mannerisms							
TG	92.38	82.88	9.50	10.70	[.55, 18.45]	0.89	.040
CG	95.00	94.80	0.20	7.22	[-8.77, 9.17]	0.03	.954

Note. IG = intervention group, CG = comparison group. Effect size is Cohen's *d*. *p* = Two-tailed. Scores for subscales are reversed.

Significance levels for the total score reached significance ($p = .012$) for the intervention group between pre and posttest questionnaire scores. The intervention group mean decreased ($\Delta M = 6.00$) from pretest to posttest and more importantly a very large effect size was observed ($d = 1.18$). The group mean for the subscale social communication decreased ($\Delta M = 7.25$), this result coupled with a significance level being reached (p = .003), shows there was a significant difference within the intervention group. Additionally, a very large effect size was also revealed (d = 1.57), further supporting that the hypothesis.

For the subscale, autistic mannerisms, the intervention group mean also decreased (ΔM = 9.50), and displayed a significance level of p = .040. A large effect size was also revealed (d = 0.89), further supporting the hypothesis.

Child Behavior Checklist

A paired t-test did not reveal significant pre and posttest differences within intervention groups for the total scale, as well as subscales.

Table 4: Chile Behavior Checklist Results

Scales	Pretest Mean	Posttest Mean	Δ Mean	*SD*	95% *CI*	Effect size	*p*
Total scale:							
TG	62.00	62.25	1.75	8.66	[-5.49, 8.99]	0.20	.596
CG	69.40	66.20	3.20	7.56	[-6.19, 12.59]	0.42	.398
Syndrome scales:							
Internalizing							
TG	54.13	57.20	-3.13	18.92	[-18.95, 12.70]	-0.17	.655
CG	61.20	61.40	-0.20	2.68	[-3.53, 3.13]	-0.07	.876
Externalizing							
TG	56.63	55.38	1.25	9.22	[-6.46, 8.96]	0.14	.713
CG	57.40	62.80	-5.40	15.53	[-24.69, -13.89]	-0.35	.480
Competence							
TG	31.38	30.50	0.88	9.31	[-6.91, 8.66]	0.09	.798
CG	27.20	29.60	-2.40	2.79	[-5.87, 1.07	-0.86	.127

Note. IG = intervention group, CG = comparison group. Effect Size is Cohen's d. 95% Confidence Interval of the difference. p = Two-tailed.

Sensory Profile

A paired *t*-test showed significant differences from pre to posttest scores within the intervention group for the multisensory processing subscale. These results are summarized in Table 5. The comparison group showed significant differences from pre to posttest within the emotional and social responses and the sensory seeking subfactor scale.

Table 5: Sensory Profile Results: Sensory Modulation

Sensory modulation	Pretest Mean	Posttest Mean	Δ Mean	*SD*	95% *CI*	Effect size	*p*
Multisensory processing							
IG	20.38	23.28	-3.00	3.07	[-5.57, -.43]	-0.98	.028
CG	21.20	21.60	-0.40	3.05	[-4.19, 3.39]	-1.31	.784

Note. IG = intervention group, CG = comparison group. Effect Size is Cohen's *d*. 95% Confidence Interval of the difference. *p* = Two-tailed.

Table 6: Sensory Profile Results: Behavioral and Emotional Responses

Behavioral and emotional responses	Pretest Mean	Posttest Mean	Δ Mean	*SD*	95% *CI*	Effect size	*p*
Behavioral and emotional							
IG	61.00	59.13	1.88	14.30	[-10.0, 13.84]	0.13	.722
CG	59.40	63.80	-4.40	1.82	[-6.66, -2.14]	-2.42	.006

Note. IG = intervention group, CG = comparison group. Effect Size is Cohen's *d*. 95% Confidence Interval of the difference. *p* = Two-tailed.

Conclusion

The future of hippotherapy relies on it being considered among objectively investigated medical and therapeutic treatments, not among the miraculous and mysterious healing practices. If most hippotherapy practitioners and researchers would agree to accept and follow a scientific approach in its evaluation, more sophisticated research challenges would arise and could be addressed.

In many fields of science, especially where qualitative evaluation or quantitative measurement is required, much effort is put towards developing common guidelines or directives concerning the conduct of an experiment. This process, called standardization, is also needed in hippotherapy, in accordance with the rules of science. The more issues we have standardized, the more common language we speak and thus, the more informative the dissipation of our knowledge. It is up to practitioners and stakeholders to decide whether hippotherapy is a mystical art of healing or a solid art of rehabilitation. It is also up to the field of hippotherapy to decide how non-medical and medical professions view this treatment strategy.

I, as a hippotherapy practitioner, would like to encourage you to believe in experiments, numbers and mathematics. Why? Because it is a scientifically reliable way of propagating hippotherapy. If our studies are properly designed and conducted in accordance with a scientific approach, the evidence presented will be very hard to deny.

Instead of trying to attach spectacular physical theories to hippotherapy, we could carefully perform a series of laborious quantitative measurements (or at least qualitative evaluation) and process the data statistically in order to be able to say that this treatment strategy is more effective than traditional physical therapy in cases of muscle contracture, for instance, with a probability of x and with statistical significance. This will be received differently by the medical community than something like "because of the impact of energy fields, the horse takes bad energy from the patient and heals him/her".

REFERENCES

Achenbach, T. M. (1991). Manual for the Child Behavior Checklist (4-18 and 1991 Profile). Burlington: University of Vermont, Department of Psychiatry.

All, A. C., Loving, G. L., & Crane, L. L. (1999). Animals, horseback riding, and implications for rehabilitation therapy. Journal of Rehabilitation, 65(3), 49-57.

American Psychiatric Association. ((2013). Diagnostic and Statistical Manual of Mental Disorders (5th ed.). Washington, DC.

Bass, M. M., Duchowny, C. A., & Llabre, M. M. (2009). The effect of therapeutic horseback riding on social functioning in children with autism. Journal of Autism and Developmental Disorders, 39, 1261-1267.

Bauman, M. L., & Kemper, T. L. (1995). Neuroanatomic observation of the brain in autism. In M. L. Bauman & T. L. Kemper (Eds.), The Neurology of Autism (pp. 119-145). Baltimore, MD: John Hopkins University Press.

Biery, M. J. (1985). Riding and the handicapped. Veterinary Clinic North America: Small Animal Practice, 15, 345-354

Bliss, B. (1997). Therapeutic horseback riding? Rehabilitation Nursing, 60, 69-70.

Cole, A. (1992). Symbol and metaphor in therapeutic riding. NARHA News, 5(1), 8.

Constantino, J. N. (2002). The Social Responsiveness Scale. Los Angeles, CA: Western Psychological Services.

Dunn, W. (2006). The Sensory Profile School Companion. San Antonio, TX: Pearson.

Dunn, W. (1999). The Sensory Profile: Examiner's Manual. San Antonio, TX: Psychological Corporation.

Freeman, B., Ritvo, E., Needleman, R., & Yokota, A. (1985). The stability of cognitive and linguistic parameters in autism: A 5-year study. Journal of the Academy of Child Psychiatry, 24, 290-311.

Frewin, K., & Gardiner, B. (2005). New age or old sage? A review of equine assisted psychotherapy. Australian Journal of Counseling Psychology, 6, 1-15.

Gilliam, J. E. (2006). The Gilliam Autism Rating Scale – Second Edition. Austin, TX: Pro-Ed, Inc.

Gray, A. (2007). The Effects of Therapeutic Horseback Riding with Autistic Children (Doctoral dissertation). Available from ProQuest Dissertations and Theses Database. (UMI No. 3288822)

Haehl, V., Giuliani, C., & Lewis, C. (1999). Influence of hippotherapy on the kinematics and functional performance of two children with cerebral palsy. Pediatric Physical Therapy, 11(2), 89-101.

Horses and Humans Research Foundation. (2010). Glossary of terms for writers and researchers. Retrieved from http://horsesandhumans.org/Research_Terminology.html

Kaiser, L., Spence, L. J., Lavergne, A. G., & Vanden Bosch, K. L. (2004). Can a week of therapeutic riding make a difference? A pilot study. Anthrozoös, 17(1), 63-72.

Keller, F., & Persico, A. M. (2003). The neurobiological context of autism. Molecular Neurobiology, 28(1), 1-22.

Land, G., Errington-Povalac, E., & Paul, S. (2001). The effects of therapeutic riding on sitting posture in individuals with disabilities. Occupational Therapy in Health Care, 14(1), 1-12.

MacKinnon, J. R., Noh, S., Laliberte, D., Lariviere, J., & Allan, D. E. (1995). Therapeutic horseback riding: A review of the literature. Physical and Occupational Therapy in Pediatrics, 15(1), 1-14.

Martin, F., & Farnum, J. (2002). Animal-assisted therapy for children with pervasive disorders. Western Journal of Nursing Research, 24(6), 657-670.

Meyer, L., Fox, A., Schermer, A., Ketelsen, D., Montan, N., Maley, K., & Cole, D. (1987). The effects of teacher intrusion on social play interactions between children with autism and their non-handicapped peers. Journal of Autism and Developmental Disorder, 17, 315-332.

Nimer, J., & Lundahl, B. (2007). Animal-assisted therapy: A meta-analysis. Anthrozoös, 20(3), 225-238.

Professional Association of Therapeutic Horsemanship. (n.d.). About PATH Intl. (n.d.). Retrieved from http://www.pathintl.org/about-path-intl/about-path-intl

Pierce, K., & Courchesn e, E. (2001). Evidence for a cerebellar role in reduced exploration and stereotyped behavior in autism. Biological Psychiatry, 49(8), 655-664.

Redefer, L. A., & Goodman, J. F. (1989). Brief report: Pet-facilitated therapy with autistic children. Journal of Autism and Developmental Disorders, 19(3), 461-467.

Roder, P. M. (2000). The early origins of autism. Scientific American, 282(2), 56-63.

Rosenzweig, M. R., Breedlove, S. M., & Leiman, A. L. (2002). Biological Psychology. Sunderland, MA: Sinauer Associates.

Rutter, M., Bailey, A., & Lord, C. (2003). The Social Communication Questionnaire Manual. Los Angeles, CA: Western Psychological Services.

Walsh, F. (2009). Human-animal bonds I: The relational significance of companion animals. Family Process, 48, 462-480.

Ward, S., Whalon, K., Rusnak, K., Wendell, K., & Paschalll, N. (2013). The association between therapeutic horseback riding and the social communication and sensory reaction of children with autism. Journal of Autism & Developmental Disorders, 43(9), 2190-2198

APPENDICES

APPENDIX A: TREATMENT GROUP INFORMED CONSENT FORM TO PARTICIPATE IN A RESEARCH STUDY

To: Research Study Participant's

From: Cherié Page, Primary Researcher

You are invited to participate in a research study that is being conducted to contribute to the exploration of the benefits of therapeutic riding for children has who have been diagnosed with Autistic Spectrum Disorder and to the field of transpersonal psychology. Your participation may contribute to your own understanding of the impact your particular experience has had in your life and your child's.

Your involvement will consist of: filling out four different questionnaires (10-15 minutes each) at the beginning and at the end of each period of treatment, and participating in an interview at the beginning of the study wherein you will be asked about your experience with your child (approximately 45 minutes). Minimal physical effort will be involved. The interview will be audio-recorded at DreamPower Horsemanship or other mutually agreed upon convenient location. The audio-recording will be transcribed and themes and patterns that emerge from the text will be included and published in the final research study. If geographical distance is a factor, the interview may be conducted via telephone with all telephone expenses paid for by me.

To assure your privacy and the confidentiality of your interview responses, you will be asked to choose a pseudonym that will be used in the final research study. Please write the pseudonym on the space provided at the end of this form. I will keep all interview files (including audio file) in a locked cabinet. If a transcriber is used, the transcriber will be required to sign a transcriber confidentiality agreement. As an added precaution, I will alter any information which might otherwise identify you as a participant in this research study.

Before signing this consent form and participating in the study, please consider the possibility that discussing your personal life experience may bring up memories or uncomfortable feelings. If at any time you have concerns or questions during the course of the interview, I will make every effort to discuss them with you and inform you of various options to resolve your concerns. In addition, you may withdraw from the study at any time without penalty or prejudice.

You may also request a written summary of the research findings by providing your mailing address. I can be reached at (xxx) xxx-xxxx. You can also contact Charlotte Lewis, Ph.D. (Chairperson of Dissertation Committee), or Fred Luskin, Ph.D. (Chairperson for the Research Ethics Committee) at the Institute of Transpersonal Psychology, 1069 E. Meadow Circle, Palo Alto, CA 94303 at 650-493-4430.

I attest that I have read and understand this consent form. Any questions I have about this research study and my participation have been answered to my satisfaction. I understand that my participation is entirely voluntary and that no pressure has been applied to encourage participation. My signature indicates my willingness to participate in this research study and to have the results published.

Write pseudonym here: ______________________________________

____ ___________________

Participant's Signature Date

Waiver of Assent:
The assent of ______________________________ (name of child/minor) was waived because of:____

Age _________ Maturity ________ Psychological state of the child _______________________

____ ___________________

Parent's Signature Date

____ ___________________

Researcher's Signature Date

Please send me a written summary of the study's pertinent findings: ___ yes ___ no

Contact Information (Please Print):
Name: ____
Address: ____
____
Phone: ____
E-mail: ____

Appendix B: Treatment Group Informed Consent Form for Participates Age 5 to 12 to Participant in a Research Study

Investigator: Cherié E. Page

I am doing a research study about how Therapeutic Riding can help children. A research study is a way to learn more about people. If you decide that you want to be part of this study, you will be asked to come once a week and do Therapeutic Riding for half an hour.

There are some things about this study you should know. These are agreeing to do Therapeutic Horseback Riding for 15 weeks and wearing a helmet to protect your head from injury.

If you do not want to be in this research study, we will tell you what other kinds of treatments there are for you.

When we are finished with this study we will write a report about what was learned. This report will not include your name or that you were in the study.

You do not have to be in this study if you do not want to be. If you decide to stop after we begin, that's okay too. Your parents know about the study too.

If you decide you want to be in this study, please sign your name.

I, ______________________________, want to be in this research study.

________________________________ ________ ________

Sign your name here Date Age

I have talked with my child about the study and feel that they understand what they will be asked to do and their rights as a study participant.

____ ________________

Parent's Signature Date

____ ________________

Researcher's Signature Date

ARTICLE FOUR

Equine-Assisted Rehabilitation for Breast Cancer Survivors

Aliza R. Hirsch, MA
The Therapeutic Riding Centre of Israel

Correspondence to:
Aliza R. Hirsch
The Therapeutic Riding Centre of Israel
Halord Street, POB 3737
Tel Mond, Israel, 40600
Email: alidavid@netvision.net.il

Support:
The Therapeutic Riding Centre of Israel helped fund the pilot program and research reported on in this article.

Acknowledgements:
I am grateful to the Therapeutic Riding Centre of Israel for their permission and funding of the pilot program and research reported in this article. I would also like to thank the Israeli Physiotherapy Society, the Israeli Society for Clinical Oncology and Radiation Therapy- ISCORT, the Israeli Cancer Association, the One in Nine Organization, the Israeli Health Ministry, Susan G. Komen, and the Israeli Equestrian Federation for their support.

ABSTRACT

Breast cancer is one of the most common malignancies that affect women. Greater awareness, early detection, and swift intervention have reduced mortality rates. Up to 80% of breast cancer survivors attain a full life expectancy (Fisher & Howell, 2010). Breast cancer treatments, however, often negatively impact upper extremity function and the overall perceived quality of an individual's daily life. Furthermore, cancer survivors may suffer diminished self-efficacy and functional deficits, along with stress-related symptoms such as post traumatic stress disorder. It is globally recognized that post-intervention physical and functional rehabilitation has been inadequate. The benefits of equine-assisted therapy (EAT) for people who cope with physical and neurological impairments have been well documented. However, most medical professionals are unaware of the potential benefits of EAT for breast cancer survivors. This paper introduces a three-phase EAT programme for breast cancer survivors and reports on its application with three clients between 2009 and 2014. Finally, a research proposal is introduced with the following research question: how does EAT impact the range of motion (ROM) and self-efficacy of breast cancer survivors.

KEYWORDS

Breast Cancer; Equine-Assisted Therapy; Quality of Life; Range of Motion; Rehabilitation.

BACKGROUND

EQUINE-ASSISTED REHABILITATION FOR BREAST CANCER SURVIVORS

Breast cancer is one of the most common malignancies affecting women. Awareness, early detection, and swift intervention have significantly reduced mortality rates (World Health Organization, 2012). Schmitz et al. (2012) noted: "for most women in developed countries, breast cancer has become a survivable chronic disease" (p. 2191).

Israel ranks fifth worldwide in the incidence of breast cancer (Israel Cancer Association, 2012). One in eight women are at risk. The disease is on the rise among Jewish Israeli women, but mortality rates have declined. At present, 90% of all cases can be cured if the disease is detected early. In Israel, breast cancer age and standardized mortality rates have dropped from 22.6% in 1998 to 17.4% in 2009 (Israel Cancer Association, 2012).

BREAST CANCER TREATMENT AND THE NEED FOR REHABILITATION

Breast cancer treatments typically involve surgery, chemotherapy, and radiation therapy. These therapies have physiological effects on body function and structure, such as overall fatigue, pain, lymphedema, restricted range of motion (ROM), peripheral neuropathy, and osteoporosis (Rock & DeMichelle, 2003), along with stress-related symptoms and low self-efficacy, all of which lead to limitations and restrictions in the performance of daily activities (Fisher & Howell, 2010).

Impairments related to breast cancer often go unrecognized. Untreated, they negatively impact function, leading to decreased tolerance to activity, and limitations and restrictions in activities of daily living (ADL), that is, the routine activities of eating, bathing, dressing, toileting, walking, and continence (Schmitz et al., 2012). Physical and functional recovery has been relatively neglected.

Osteopenia and osteoporosis significantly affect quality of life (QOL) (National Institute of Arthritis, Musculoskeletal and Skin Diseases, 2012) and contribute to reduced ADL and ROM, limit independence, raise anxiety, and lower self-image. These sequalae fit in the models of International Classification of Functioning (ICF) and Rehabilitation Empowerment presented later. Al Amri and Sadat-Ali (2009) confirmed that "there is a high prevalence of osteopenia and osteoporosis in patients who were younger than 50 years" (p. 331). Ramaswamy and Shapiro (2003) asserted that since "many women with breast cancer will be long-term survivors, increasing recognition of maintaining skeletal health is important" (p. 763).

The natural emphasis on survival somewhat overshadows the long-term treatment-related medical and psychological sequelae. Today, nearly 80% of breast cancer survivors can attain full life expectancy and they should do so with full functional capacity (Fisher & Howell, 2010).

Coleman (2006) suggested that appropriate rehabilitation can maximize the functional status of breast cancer patients, reduce the occurrence of intrusive thoughts, and address psychological and vocational problems, leading to improved ADL and QOL. In a lecture presented at Rush University Medical Center, Professor Sheila A. Dugan (2010) described the role of rehabilitation in breast cancer treatment and recovery as "the gift and opportunity to 're-build' the posture, trunk, pelvic floor, shoulder, lower back, fitness level, and self awareness, in order to enable survivors to become thrivers" (slide 33). She further emphasized that "no other body part is amputated or reconstructed without physical therapy as the first line of the rehabilitation process, except the breast" (slide 10).

Breast Cancer and Post Traumatic Stress Disorder

The emotional impact of a cancer diagnosis, symptoms, treatment, and related issues can be severe. Breast cancer survivors may develop a distorted body image or self-concept, change in family relations, and/or intrusive thoughts about recurrence or death, all of which can be relevant factors in decreased QOL, self-efficacy, belief in one's ability to overcome hardships, and empowerment, a process by which individuals attain power through access to various resources that give them control of their lives (Fisher & Howell, 2010). In June 2010, The US Cancer Institute reported that stress and trauma-related symptoms were being seen in breast cancer survivors, resembling those who experienced traumatic events (Cordova et al., 2000).

Women diagnosed with breast cancer should be evaluated and considered at risk of developing post traumatic stress disorder (PTSD). Under the diagnostic criteria for PTSD in the *Diagnostic and Statistical Manual of Mental Disorders* (5th ed.; *DSM-V*; American Psychiatric Association, 2013), an example of a traumatic event is being diagnosed with a life-threatening disease.

Diagnostic Criteria and Characteristics

PTSD was initially characterized as an anxiety disorder that developed in response to a severe trauma in which an individual experienced, witnessed, or was confronted by actual or threatened death, injury, or loss of physical integrity of self or others. The *Diagnostic and Statistical Manual of Mental Disorders* (4th ed.; *DSM-IV*; American Psychiatric Association, 1994) added that diagnosis for a life-threatening illness qualifies as a stressful event.

The first known study of the prevalence of PTSD among breast cancer survivors using the *DSM-IV* criteria included 27 patients, each with a minimum of three years of post-cancer treatment. The results indicated a prevalence rate of 40% for acute PTSD and 22% for chronic PTSD (symptoms lasting more than three months) (Andrykowski & Cordova, 1998). More recent research revealed that two to three months after diagnosis, 23% of breast cancer survivors met the criteria for acute PTSD (Vin-Raviv et al., 2013)..

Breast Cancer and Self-Efficacy

Breast cancer treatments negatively impact QOL and upper body function. Survivors may experience diminished self-efficacy related to functional deficits that result from their physical limitations. The International Classification of Functioning (ICF) provides a framework for rehabilitation practitioners to address physical and psychological impairments, activity limitations, and participation restrictions, thus shifting the focus from diagnosis to function (see Figure 1).

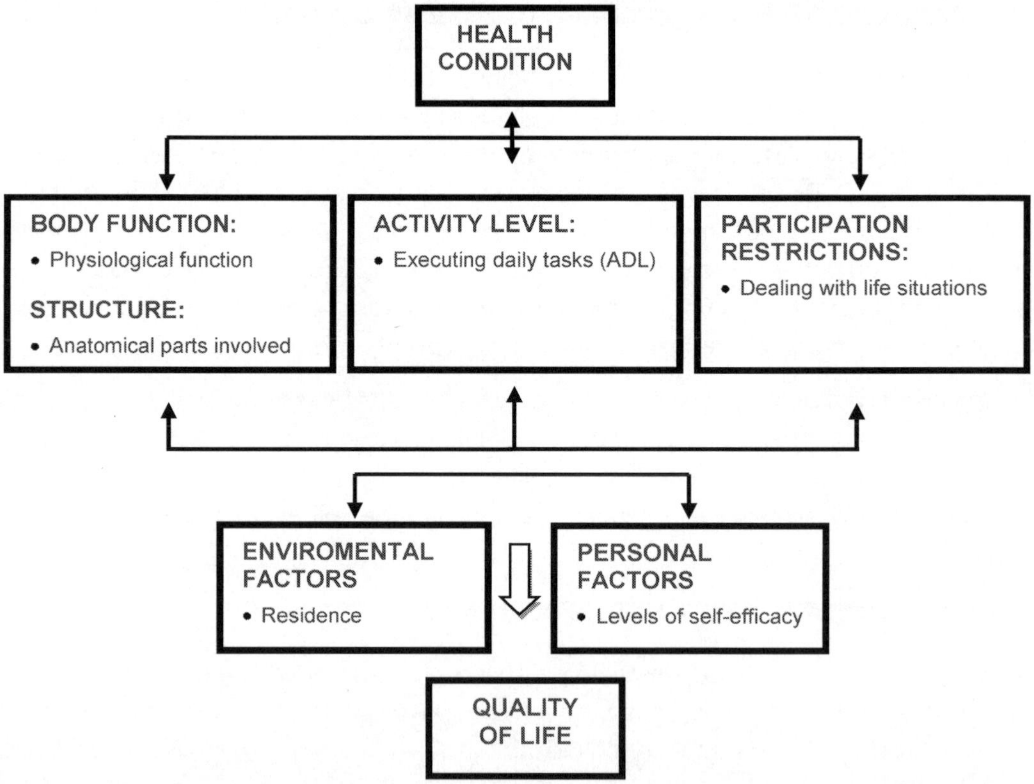

Figure 1. The International Classification of Functioning (ICF) Model

Swisher et al. (2010) concluded that upper extremity impairments were seen up to 30 years following breast cancer treatment, resulting in activity limitation and participation restrictions. This relationship between impairments, activity limitations, and subsequent participation restrictions for breast cancer survivors is illustrated in Figure 2.

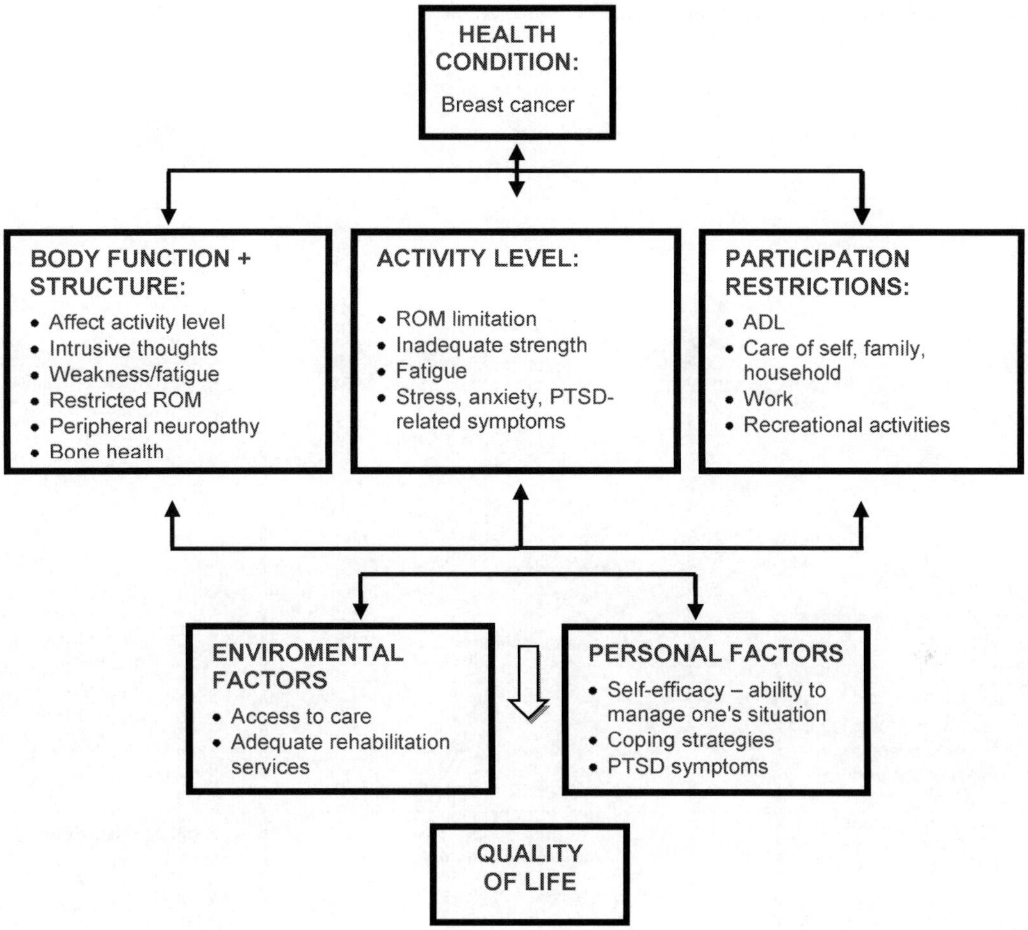

Figure 2. Application of the ICF Model to Breast Cancer

Patient outcomes may be improved by fostering self-efficacy through empowerment, that is, the belief in one's own abilities to exercise influence over the events that affects one's life (Bandura, 1998). Fisher and Howell (2010) theorized that rehabilitation practitioners could employ empowerment theory to help improve the self-efficacy of breast cancer survivors, by increasing the knowledge, skills, and resources that are needed to increase control over social and environmental factors in their life. In their study of how a combination of the ICF model and theories of self-efficacy and empowerment might improve the rehabilitation outcomes for breast cancer survivors, Fisher and Howell (2010) found that "higher levels of self-efficacy among breast cancer survivors may translate into improved functioning and QOL" (p. 19). Their model is presented in Figure 3.

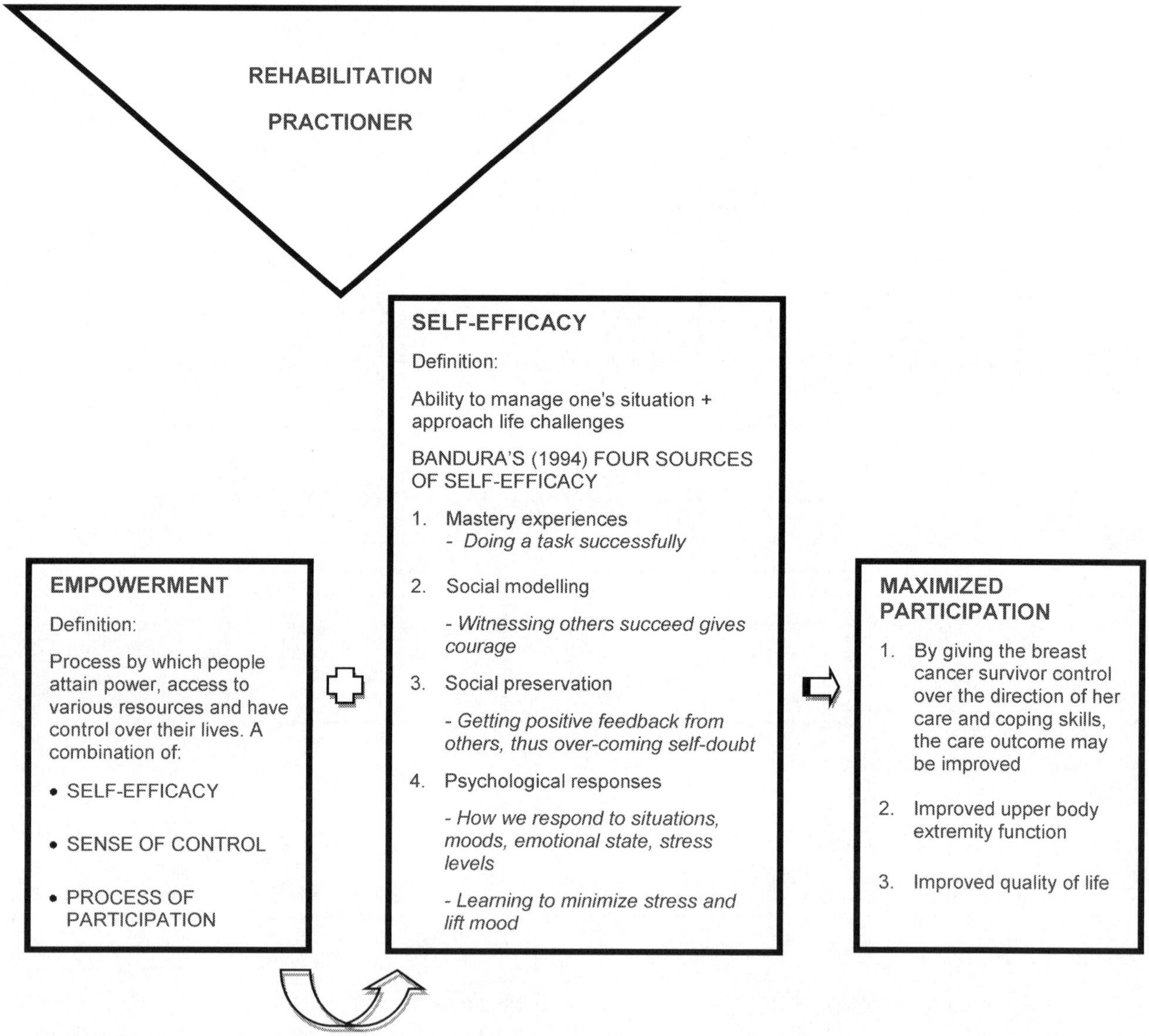

Figure 3. Fisher & Howell's Model for Rehabilitation Through Empowerment

As a basis for creating an equine-assisted therapy program for cancer survivors, I developed a model in which I replaced the "rehabilitation practitioner" with a horse. Thus the aim of the program was to raise the breast cancer survivor's self-efficacy levels using equine-assisted therapy. (Figure 4)

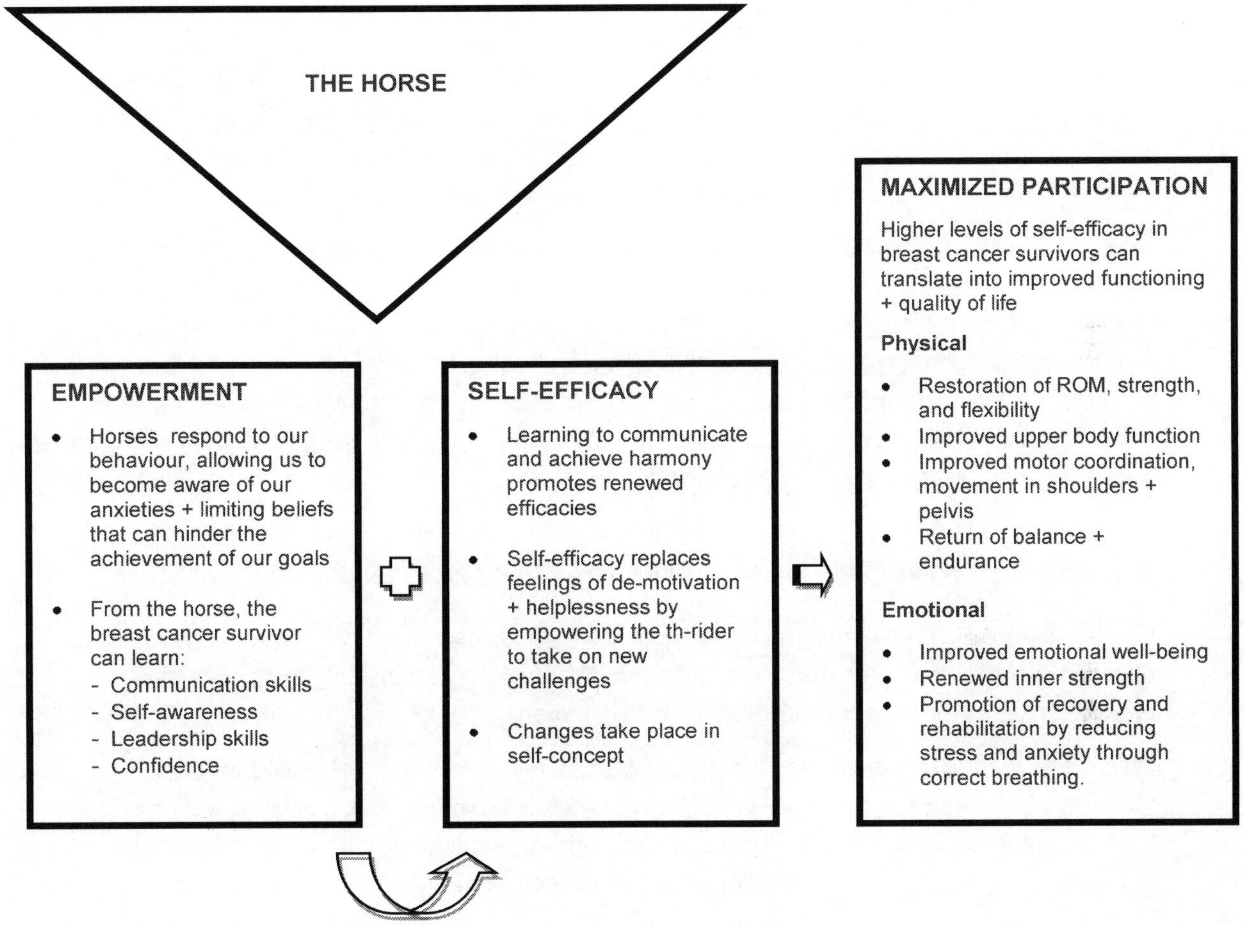

Figure 4. The Israeli Model for EAT Rehabilitation of Breast Cancer Survivors

Therapeutic Riding In Israel

The Therapeutic Riding Centre of Israel (TRCI), a therapeutic riding center and training center for therapeutic riding teachers, including internship, was founded in 1986. It cooperates with the EU Pegasus educational program and the Animal Assisted Therapies (AAT) department of Hogeschool in Ghent, Belgium.

The Israeli Ministry of Sport requires that therapeutic riding teachers hold two certifications: that of riding instructor (English or Western style) and that of therapeutic riding instructor. The riding instructor's certificate is awarded after a nine-month program, covering equine management, horsemanship, stable management, equine behaviour, equine health, instruction, and teaching methodology. The English-style course is based on the British Horse Society's Stage 2 and Preliminary Teaching Test; the Western style course is based on the basic program of the American Riding Instructors Association. All instructors and coaches are trained at an international level and are members of the Israeli Equestrian Federation.

The therapeutic riding instructor's certificate is awarded after a two-year program that includes studies in human anatomy, physiology, neurology, psychology, child development and disabilities, reading evaluations, therapy, and lesson planning. The final year is devoted to internship at a recognized ranch in Israel, working with a high-level instructor and social worker. Students are required to work with three physically and three emotionally challenged riders over a course of ten months, including design of individual therapy plans. At the end of their studies, they present these six cases to the examination board for review.

Equine-Assisted Therapy in Breast Cancer Rehabilitation

Equine-assisted therapy (EAT) incorporates equine activities and/or the equine environment. The rehabilitative goals are designed to meet the individual client's needs and the medical professional's standards of practice (PATH International, 2014).

Physically, the horse's movement at a walk creates three-dimensional movement in the rider's torso, similar to that produced when a human walks. This movement creates physical and sensory input that is variable, rhythmic, and repetitive (Uchiyama, Ohtani, & Ohta, 2011). Interaction with the horse creates opportunities for collaboration and feedback both from the facilitator and the horse, experiences of task-completion, and a motivational space to practice acceptance and awareness (Gergely, 2012). This promotes self-efficacy.

In light of these features, EAT can help breast cancer survivors in numerous ways.

- The constant repetition of the horse's stride and absorption of this rhythm into the rider's pelvis causes ligaments to relax and become suppler. It improves vertical and horizontal balance and muscle tone, as well as hip and shoulder flexibility, promoting healthy walking (Borzo, 2002).
- Exercise on horseback improves circulation and endurance (Borzo, 2002).
- Development of correct nasal diaphragmatic breathing restores mobility in the thoracic area, thus reducing stress and restoring upright posture. This is promoted by the horse's breathing, as it is a nasal breather, anatomically unable to breathe through the mouth (Riegel & Hakola, 2000).

- Equine-assisted therapy may help improve neurological functioning and sensory processing (American Hippotherapy Association, n.d), which can be generalized to a wide range of daily activities
- Through learning to communicate with the horse, the rider may develop a sense of self-efficacy and well being.

In Israel, EAT has been offered for many years to wounded veterans and terror victims (Shkedi, 2012). However, most professionals in the relevant fields are unaware of the potential benefits of EAT for breast cancer survivors.

A Three-Phase Equine-Assisted Therapy Program for Breast Cancer Survivors

In light of the indications of its potential, I developed and implemented a three-phase EAT rehabilitation program for breast cancer survivors. In the description of the program and the cases, I refer to the breast cancer survivor as a therapeutic rider.

Phase I. The program begins with an introductory phase designed to build trust and communication between the therapeutic rider, instructor, and horse on the ground and in the stable. Phase I consists of 8 sessions. The first sessions are dedicated to introducing the program, the building of trust and communication, meeting with the social worker, presenting medical records, and completing questionnaires to help assess the therapeutic rider's functional and emotional state pre-intervention. The therapeutic rider is encouraged to keep a riding journal of personal reflections. At this point, the instructor asks the therapeutic rider to voice her expectations and together they build a joint therapy plan.

The therapeutic rider gets to know the horse in the stables and in the riding arena, where grooming, learning the horse's body language, and overcoming fear of the horse are used to arouse the therapeutic rider's senses and sensory reactions. Emphasis is placed on the instructor's observation of the therapeutic rider's ROM from the ground and building trust between the therapeutic rider, horse, and instructor.

In addition, this phase features an introduction to diaphragmatic breathing, using the horse as a coach. The benefit of diaphragmatic breathing to cancer survivors has been demonstrated by Fialka et al. (2003). As obligate nasal breathers, horses are valuable teachers. At rest, or after minor exertion, the horse breathes only 8-14 breathes per minute. Synchronization of the therapeutic rider's breathing with that of the horse can decrease thoracic shallow breathing, or hypoventilation, which is a symptom of anxiety.

The other sessions in this phase are dedicated to the practice of nasal diaphragmatic breathing while mounted and an introduction to the horse's movement, leading to work on improving core strength and stability.

The overall therapy goals for Phase I are:

- to develop correct nasal diaphragmatic breathing;
- to improve the therapeutic rider's ROM and flexibility; and
- to build self-confidence.

PHASE II. After observation and assessment of the therapeutic rider in Phase I, Phase II begins with the development of an individual therapeutic program based on the therapeutic rider's needs and abilities. The emphasis is physical development, consisting of exercises designed to restore ROM and diaphragmatic breathing.

The horse is chosen to suit the therapeutic rider's build and needs, in terms of size and temperament. The work in this phase is done using a two-handled vaulting surcingle, pad, and sheepskin. This is particularly helpful in the colder months.

Phase II consists of 20-24 sessions, which include a combination of yoga, correct breathing, and physical exercises. In addition, the instructor encourages concentration on the here and now, together with self-awareness and self-acceptance based on sensing acceptance from the horse.

The overall therapy goals for Phase II are:

- to facilitate and improve the therapeutic rider's overall condition and stamina for faster recovery, maximize body strength, and improve and maintain muscle tone and bone density;
- to restore and learn to maintain ROM, with focus on shoulders, chest, upper and lower back, rotator cuff, abdominals, and lower extremities;
- to restore strength and flexibility to the involved area of the body, joints, and muscle groups that have been weakened by surgery and treatments;
- to retrain torso muscles for improved posture;
- to develop correct abdominal breathing to help restore mobility to the thoracic area, reducing stress and anxiety and allowing more freedom of movement in the diaphragm and lungs; and
- to develop self-awareness and self-acceptance.

PHASE III. The third and final phase of the program includes continued improvement of ROM through ground work, grooming, and stable work, and introduction to conventional saddle-riding skills, as well as continued development of the interpersonal and intrapersonal skills discussed earlier.

PILOT STUDY: OBSERVATION OF EAT WITH THREE CANCER SURVIVORS

Between 2009 and 2014, three Israeli women aged 35-70 participated in EAT as part of their rehabilitation program following treatment of breast and other cancers. All had undergone radical surgery, chemotherapy, and radiation therapy, and were taking the hormone-therapy drug, Tamoxifen. The HMOs to which the women belonged offered them each three months of physical therapy. Later, they participated in equine-assisted therapy at TRCI, based on the model that I developed. The horses and tack were chosen to suit each participant.

The following descriptions of the three cases, including initial presentation and the results of their EAT, are based on my observations, video recordings of sessions, and the riding journals kept by the participants.

PARTICIPANT 1

Presentation: A 51-year-old Israeli woman, married. A book restorer by profession, currently working as a therapeutic riding instructor.

Medical presentation:

- 1995 Stage III breast cancer; radical mastectomy (while pregnant), chemotherapy, and radiation therapy.
- 1997 Ovarian cancer; radical hysterectomy and chemotherapy.
- 2005-2007 Cerebral vasculitis; CVA, hemiparesis right side. (The CVA caused damage to frontal, parietal, and cerebellum areas of the brain; the right side of the body was affected: 60% hearing loss in right ear, 45% vision lost in right eye.)
- 2008 Severe osteoporosis.
- 2009 Malignant choroidal melanoma.

Riding history:

- Rode as a child in the UK.
- 2009–present. Still rides twice a week with an advanced dressage instructor.

Questionnaires used:

- TRCI medical questionnaire, completed by every rider at the center.
- Bi-annual bone mineral density (BMD) scans

Treatments offered by HMO:

- Three months of physical therapy, during period of chemotherapy. The EAT treatment of Participant 1 and the results measured are presented in Table 1.

TABLE 1: PARTICIPANT 1, 2009–2014

Therapy goals	Horses used	Results
To rediscover skills related to riding	Shiny, 16 year old Haflinger; exceptional conformation, a true breathing coach	• The client reported feeling pleasure while riding.
To improve ROM for ADL • Improve balance/stability • Strengthen shoulders • Strengthen upper body	Matty, a 15-year-old Appaloosa	• Extensive vaulting work produced excellent results in terms of balance and stability. • ADL functioning reached near-normal level. • Major improvement in ROM, with a flexible upper body in the saddle and in ADL. After two years, she was able to reverse her car into a parking space.
To improve bone density	Ginger, a 9 year old local thoroughbred; a rescued horse.	BMD scores 2007: 0.889; 2014: 1.108 T-Scores 2007: -2.5; 2014: -0.9

Therapy goals	Horses used	Results
To promote empowerment and self-efficacy	Black Star, a 14 year old Hanoverian thoroughbred; formerly a police horse	• Anxiety: Through breathing and riding, she acquired tools to calm herself. This was especially apparent before oncological visits and receiving test results. In times of stress, she continued to return to VL. • Self-esteem: She began to pay more attention to how she dressed and wear make-up. She was better able to regulate her emotions, in and out of the arena. She gained self-esteem: "I am not a bad person, I didn't deserve to get cancer." She came to love herself: "Take me as I am."
To promote empowerment and self-efficacy		• Self-efficacy: She learned to communicate better and achieve inner harmony, which promoted self-efficacy. • Empowerment: A sense of "yes, I can" replaced lack of motivation and helplessness. In 2009, she returned to college and became a therapeutic riding instructor. She realized it was ok to "fail", undertook new challenges, and fulfilled dreams.
To develop correct breathing in order to relieve anxiety, build upper body strength, and develop correct walking habits	Black Star, a 14 year old Hanoverian thoroughbred; formerly a police horse	• Correct breathing became natural, in and outside of the riding arena, reducing anxiety and helping the work on correct posture.

Therapy goals	Horses used	Results
To relieve peripheral neuropathy in hands		• A major improvement was achieved over time. • She attained the ability to hold reins and maintain contact. By 2014, she was riding first level dressage. • In ADL, she hardly dropped anything anymore, and used breathing to control the symptoms of neuropathy.
Improvement of cognitive skills • Strengthen motor function-right side • Sequencing • Memory		A major cognitive improvement was apparent: • She was able to remember a dressage test. • After 3 years, she was able to ride a 20-meter circle on the right rein. • In ADL, she was able to go shopping and bring everything home

Rider's summation: "For me, the touch, smell, and warmth of the horse are my strength, giving me courage and hope. Where I doubted, he showed me the way, and taught me that I am most certainly capable. Who would have thought that I would become a therapeutic riding instructor? I have been blessed with a second chance, and for that I am so grateful."

PARTICIPANT 2

Presentation: A 41-year-old Israeli woman, married. A social worker by profession.

Medical presentation:

- 2008 Stage III breast cancer; lumpectomy, chemotherapy, and radiation therapy; frozen shoulder on right, resulting from the radiation

Riding history:

- She had never ridden before.

Questionnaires used:

- TRCI medical questionnaire

- Personal riding journal

Treatments offered by HMO:

- Participant 2 received three months physical therapy, which did not help at all and did not relieve her frozen shoulder. Due to funding constraints, her EAT therapy period was very short, consisting of only 6 sessions. Therefore, very clear and specific goals were set. The EAT treatment of Participant 2 and the results measured are presented in Table 2.

TABLE 2: PARTICIPANT 2, NOVEMBER 2011-DECEMBER 2011 (6 SESSIONS)

Therapy goals	Horses used	Results
To give the participant pleasure	Shiny, 16 year old Haflinger.	• The client enjoyed the riding experience.
To improve ROM for ADL • Remove energy blockers so she could allow herself to work on ROM • Strengthen/rebalance right shoulder • Relieve tension in lower back/lumbar area		• ADL functioning: After just 6 sessions, she was able to get in and out of her car, pick up her young daughter for a hug, and close her bra unassisted. • ROM: Energy channels were freed, especially in the right shoulder. • She gained the ability to lift her right arm after 4 sessions. • The unblocking of energy and release of right shoulder enabled her to lift her head, to look the world head on, and allow herself to feel balanced. This was very significant. • An encouraging improvement in her back: greater ROM and flexibility.
Goals on the emotional level • To deal with anxiety • To booster self-confidence, self-esteem, and confidence in body • To encourage living in the present	Shiny, 16 year old Haflinger	• Anxiety: Although the anxiety was constant, breathing exercises helped relieve it. • After three sessions, she began using the word "cancer"; this was the first time she had discussed her disease outside of the oncology department.

Therapy goals	Horses used	Results
• To allow acknowledgement of cancer without fear		
To develop correct breathing in order to • Relieve anxiety • Foster a positive physical self-image		• She learned and adopted the breathing technique. • This was very effective in relieving anxiety and enabling work on ROM.

Rider's summation: "This kind of work [is] very suitable for women recovering from breast cancer. It is so important to increase ROM, to reconnect to the body... It is wonderful to "dump" my worries onto the horse. I was amazed that after such a short time, I could see that Shiny could sense my physical and emotional state, even though I finished my treatments so long ago." (translated from Hebrew by author).

PARTICIPANT 3

Presentation: A 69-year-old Israeli woman, married. A retired school principal. She was the only participant who had not had breast cancer.

Medical presentation

- 2010 Endometrial cancer. Underwent a hysterectomy (LAVH).
- 2012 Diagnosis of adenocarcinoma; subsequent cholecystectomy. Underwent optimal debulking and chemotherapy.
- 2012 Post-operative hernia repair.
- 2013 Level IV spondylolithesis (displacement of vertebrae), L4-L5.

Riding history

- She had never ridden before.

Questionnaires used

- TRCI medical questionnaire
- Personal riding journal

Treatments offered by HMO

- Participant 3 received no rehabilitation therapy or physical therapy from her HMO. The EAT treatment and the results measured are presented in Table 3.

TABLE 3: PARTICIPANT 3, 2013–PRESENT

Therapy Goals	Horses used	Results
To relieve peripheral neuropathy in feet	Punk, a 15-year-old Norwegian Fjord (with VL + thick blanket)	• After working barefoot for 3 months and then continuing work for another 7 months, she was able to step on her whole foot and walk with more stability. • Self-confidence increased. • Neuropathy was hardly evident.
To give the participant an experience of fun and pleasure	Graham, a 17-year-old Danish Warmblood (with VL + thick blanket)	• The horse provided her with physical and emotional grounding. • She began looking forward to her riding lessons. • She invited her husband to watch. • During vacations, she invited her grandchildren, who also rode with her • She began laughing aloud in response to accomplishments. • She giggled and asked for more time when she trotted. • She allowed the mention of cancer.
To strengthen the participant's upper body. Areas to be worked on: • Body alignment • Lower back • Shoulders	Tomi, a 6-year-old Tinker, with a saddle	• Using a saddle gave the participant a sense of independence. • Body alignment: Major improvement. • Shoulders: She allowed herself to open up physically and emotionally. • Lower back: The tension in her lower back was reduced.

Therapy Goals	Horses used	Results
• Abdominals		• After difficulty with diaphragmatic breathing (complaining, "How can I breathe if I don't have a diaphragm?"), she began working on it. Eventually, she realized she did have a diaphragm, she could control her breathing, and it was effective in helping her work; it then became second nature.
To improve ROM for ADL • Stability		• Improvement in centering: she sat beautifully on the horse, maintained stability when walking and at a sitting trot. • ADL: Major improvement in all aspects. • Walking. She began walking on her entire foot and upright. • She was able to get in and out of a car and was able to bring shopping home.
To develop correct breathing in order to • Relieve anxiety, including the fear she would hurt herself • Strengthen upper body • Adopt correct walking habits		• Correct breathing became second nature for her in and outside of the arena. • Her anxiety was reduced and she was able to work on correct posture.
Goals on the emotional level • To develop greater relaxation and trust in the horse • To reduce the occurrence of intrusive thoughts. • To acknowledge and talk about cancer.		• Sixteen months after the riding therapy began, she finally allowed herself to relax. She formed special bonds with each horse she rode. She spoke to Tomi, and allowed herself to feel pleasure. Taking the reins helped her gain a sense of independence, liberation, and the ability to overcome

Therapy Goals	Horses used	Results
		obstacles. • She had a much less morbid outlook. The anxiety she used to express before oncological tests was replaced by a much philosophical view. • She called cancer by its name and spoke about it. She still feared the disease, but also sometimes allowed herself to feel that she had survived it.
To develop self-efficacy • Do something on her bucket list; realize a dream. • Believe in yourself: "Yes I can." • Develop positive thinking.		• She could now check an item on her bucket list (riding a horse). • She had restored her positive outlook and began making plans for a vacation. • She renewed her wedding vows.

Rider's summation. "When I ride, [I] am free [of] illness, of worries. Yes, I can. I am capable. . . . It is amazing what a horse can bring out in its rider. Today, I can fly, raise my head, face the world, the doctors, and my disease" (translated from Hebrew by author).

Research Proposal for Equine-Assisted Therapy with Breast Cancer Survivors

The present paper summarizes the implementation of my model by means of observation of three clients who participated in the three-phase EAT program at TRCI between 2009 and 2014. This study was designed as a pilot for future comprehensive research designed to formally assess the effectiveness of the program. The following is a description of the planned research program.

Research Question

The purpose of the research will be to examine how an eight-month equine-assisted therapy program affects range of motion (ROM) and sense of self-efficacy among breast cancer survivors.

Participants and Control Group

The participants will be 20 Israeli female breast cancer survivors (post-medical treatment), between the ages of 25-70 years. The researchers will recruit participants from the breast oncology departments at the four major hospitals in Israel and through the Israel Cancer Association and One-in-Nine Organization. Half of the participants will be assigned to rehabilitation programs offered by the four major HMOs in Israel (control group) and half to the same rehabilitation programs and, in addition, equine-assisted therapy intervention (experimental group).

Setting

The equine-assisted therapy program will be conducted at the Therapeutic Riding Centre of Israel (TRCI), the largest therapeutic riding center in Israel. TRCI is run by a non-profit organization of the same name, which is dedicated to improving the lives of people with disabilities through interdisciplinary professional therapy. Using unique therapeutic and recreational methods, physical therapists, social workers, speech therapists, and psychologists, and others combine their skills with those of the equine-assisted therapists to achieve these goals. TRCI serves a multicultural population and adheres to nondiscriminatory practice. All the riding teachers that participate in the research will be trained and appropriately credentialed.

Measures

The research will be designed to assess ROM, QOL, ADL and self-efficacy levels prior to the rehabilitation programme, after Phase I (6 weeks), after Phase II (6 months), and after Phase III (8 months). A final assessment will be made three months after the intervention has ended.

A set of questionnaires will be used for this purpose, including:

1. Cancer Rehabilitation Evaluation System for Research (Schag & Heinrich, 1990).
2. General Self Efficacy Scale (GSE), in English (Schwarzer & Jerusalem, 1995) and Hebrew (Zeidner, Schwarzer, & Jerusalem, 1993).
3. European Organization for Research and Treatment of Cancer Care questionnaires and breast module (EORTC-QLQ-C30/BR23), in English (Sprangers et al., 1996) and Hebrew (EORTC Quality of Life Group, 1996).
4. Disabilities of Arm, Shoulder + Hand (DASH), in English (Hudak, Amadio, & Bombardier (1996) and Hebrew (Institute for Work & Health, 2006/2012).
5. Bone mineral density (BMD) scans.

Conclusion

Breast cancer is one of the most widespread types of cancer in women. In developed countries, it has become a survivable chronic disease. The natural emphasis in treatment is on survival, but breast cancer survivors also face many post-treatment challenges and a lengthy rehabilitation period. Nearly 80% of breast cancer survivors will attain full life expectancy and they should do so with full functional capacity (Fisher & Howell, 2010).

Equine assisted therapy is a rehabilitation option that provides both physical and psychological benefits. More research is needed to validate its effectiveness for breast cancer survivors. Another aim of the three-phase program is to inspire others to offer services to this population, and to create a format for future research on interventions of this type.

In addition, to the best of my knowledge, there is no current research on the effect of EAT on bone density. In this respect, the results for Participant 1 in this research warrant further investigation, as they are encouraging.

I hope to present the findings of this future pilot study to the Israel Ministry of Health and Israeli HMOs, showing them that this type of rehabilitation program can be effective. It is my hope that equine assisted therapy will eventually be available to all breast cancer survivors worldwide, and not only a privileged few.

References

Al Amri, A., & Sadat-Ali, M. (2009). Cancer chemotherapy-induced osteoporosis: How common is it among Saudi Arabian cancer survivors. Indian Journal of Cancer, 46(4), 331-334.

American Hippotherapy Association (n.d.). Hippotherapy as a treatment strategy. Retrieved from: http://www.americanhippotherapyassociation.org/hippotherapy/hippotherapy-as-a-treatment-strategy/

American Psychiatric Association (1994). Diagnostic and Statistical Manual of Mental Disorders. (4th ed.). Washington, DC.

American Psychiatric Association (2013). The Diagnostic and Statistical Manual of Mental Disorders. (5th ed.). Washington, DC. Andrykowski, M. A. & Cordova, M. J. (1998). Factors associated with PTSD symptoms following treatment for breast cancer: Test of the Andersen model. Journal of Traumatic Stress, 11(2), 189-203.

Bandura A. (1998). Self-efficacy. In H. Friedman (Ed.), Encyclopedia of Mental Health. San Diego, CA: Academic Press,

Borzo, G. (2002). Horse power. American Medical News, 45, 24-26.

Coleman, M. (2006). Nursing Patients with Cancer: Principles and Practice. Edinburgh Elsevier Science, 801.

Cordova, M. J., Studts, J. L., Hann, D. M., Jacobsen, P. B., & Andrykowski, M. A. (2000). Symptom structure of PTSD following breast cancer. Journal of Traumatic Stress, 13(2), 301-319.

Dugan, S. A. (2010). Rehabilitation and exercise and breast cancer. Powerpoint presentation accompanying lecture. Retrieved from: http://www.rad.rpslmc.edu/Rush.../EC_0510_ Dugan_BreastCancerRehab.pdf

EORTC Quality of Life Group (1996). QLQ-BR23 [measurement instrument, translated to Hebrew]. Retrieved from: http://site-www.eortc.be/qol/downloads/QLQC30/select.asp

Fialka-Moser, V., Crevenna, R., Korpan, M., & Quittan, M. (2003). Cancer rehabilitation: Particularly with aspects on physical impairments. Journal of Rehabilitation Medicine, 35(4), 153-162.

Fisher, M., & Howell, D. (2010). The power of empowerment: An ICF-based model to improve self-efficacy and upper extremity function of survivors of breast cancer. Rehabilitation Oncology, 28(3), 19-25.

Gergely, E. (2012). Equine-Assisted Psychotherapy: A Descriptive Study (Doctoral dissertation). Retrieved from Western Michigan University Dissertations (http://scholarworks.wmich.edu/dissertations/107).

Institute for Work & Health (2006/2012). Disabilities of the Arm, Shoulder and Hand (DASH) [Measurement instrument, translated to Hebrew]. Available from efratziv1@gmail.com

Israel, B. A., Checkoway, B., Schulz, A., & Zimmerman, M. (1994). Health education and community empowerment: Conceptualizing and measuring perceptions of individual, organizational, and community control. Health Education and Behavior, 21(2), 149-70.

Israel Cancer Association (2012). http://www.cancer.org.il

Keinan-Boker, L., Baron-Epel, O., Fishler, Y., Liphshitz, I., Barchana, M., Dichtiar, R., & Goodman, M. (2012). Breast cancer trends in Israeli Jewish and Arab women, 1996-2007. International Journal of Clinical Practice, 66(10), 948-958.

National Institute of Arthritis, Musculoskeletal and Skin Diseases (2012). What breast cancer survivors need to know about osteoporosis. Retrieved from: http://www.niams.nih.gov/Health_Info/Bone/Osteoporosis/Conditions_Behaviors/osteoporosis_breast_cancer.asp

Ogburn, A., & Borton, B. (2009). Therapeutic riding and hippotherapy: What is it and how does it work? Retrieved from: http://www.speechpathology.com/articles/therapeutic-riding-and-hippotherapy-what-1177

PATH International (2014). Learn about EAAT. Retrieved from: http://www.pathintl.org/resources-education/resources/eaat

Ramaswamy, B., & Shapiro, C. L. (2003). Osteopenia and osteoporosis in women with breast cancer. Seminars in Oncology, 30(6), 763-775.

Riegal, R. J., & Hakola, S. J. (2000). Illustrated Atlas of Clinical Equine Anatomy and Common Disorders of the Horse, Volume II. Equistar Publication: Marysville, OH.

Rock, E., & DeMichele, A. (2003). Nutritional approaches to late toxicities of adjuvant chemotherapy in breast cancer survivors. Journal of Nutrition, 133 (11, Suppl 1), 3785S-3793S.

-

Schmitz, K., Stout, N., Andrews, K., Binkley, J., & Smith, R. (2012). Prospective evaluation of physical rehabilitation needs in breast cancer survivors: A call to action. Cancer, 118(8): 2191-2200.

Schwarzer, R., & Jerusalem, M. (1995). Generalized Self-Efficacy Scale. In J. Weinman, S. Wright, & M. Johnston, Measures in health psychology: A user's portfolio. Causal and control beliefs (pp. 35-37). Windsor, UK: NFER-NELSON. Retrieved from: http://userpage.fu-berlin.de/%7Ehealth/engscal.htm

Shkedi, A. (2012). Traumatic Brain Injury and Therapeutic Riding. Israel. Available from anitashkedi@me.com

Sprangers, M. A., Groenvold, M., Arraras, J. I., Franklin, J., te Velde, A., Muller, M., ... & Aaronson, N. K. (1996). The European Organization for Research and Treatment of Cancer breast cancer-specific quality-of-life questionnaire module: First results from a three-country field study. Journal of Clinical Oncology, 14(10), 2756-2768.

Swisher, A., Davison, C., Aranda, R., Eye, D., & Erikson, M. (2010). Frequency and severity of self-reported upper extremity impairments, activity limitations and participation restrictions following breast cancer treatments. Rehabilitation Oncology, 28(1), 3-9.

Uchiyama, H., Ohtani, H., & Ohta, N. (2011). Three-dimensional analysis of horse and human gaits in therapeutic riding. Applied Animal Behavior Science, 135(4), 271-276.

Vin-Raviv, N., Hillyer, G., Hershman, D., Galea, S., Leoce, N., Bovbjerg, D., … Neugut, A. (2013). Racial disparities in posttraumatic stress after diagnosis of localized breast cancer: The BQUAL study. Journal of the National Cancer Institute, 105(8), 563–572.

World Health Organization. (2012). Breast cancer: Prevention and control. Retrieved from: http://www.who.int/cancer/detection/breastcancer/en/

Zeidner, M., Schwarzer, R., & Jerusalem, M. (1993). Hebrew adaptation of the Generation Self-Efficacy Scale [measurement instrument]. Retrieved from: http://userpage.fu-berlin.de/%7Ehealth/hebrew.ht

Made in United States
Orlando, FL
04 November 2026

90206122R00050